GROUNDWATER RESPONSE TO CHANGING

SELECTED PAPERS ON HYDROGEOLOGY

16

Series Editor: Dr. Nick S. Robins
Editor-in-Chief IAH Book Series
British Geological Survey
Wallingford, UK

INTERNATIONAL ASSOCIATION OF HYDROGEOLOGISTS

Groundwater Response to Changing Climate

Editors

Makoto Taniguchi
Research Institute for Humanity and Nature (RIHN),
Motoyama Kyoto, Japan

Ian P. Holman
Department of Natural Resources, Cranfield University, UK

 CRC Press
Taylor & Francis Group
Boca Raton   London   New York   Leiden

CRC Press is an imprint of the
Taylor & Francis Group, an **informa** business

A BALKEMA BOOK

First issued in paperback 2017

CRC Press/Balkema is an imprint of the Taylor & Francis Group, an informa business

©2010 Taylor & Francis Group, London, UK

Typeset by Vikatan Publishing Solutions (P) Ltd., Chennai, India

Published by: CRC Press/Balkema
    P.O. Box 447, 2300 AK Leiden, The Netherlands
    e-mail: Pub.NL@taylorandfrancis.com
    www.crcpress.com – www.taylorandfrancis.co.uk – www.balkema.nl

Library of Congress Cataloging-in-Publication Data

Groundwater response to changing climate / editors, Makoto Taniguchi, Ian P. Holman.
    p. cm. -- (Selected papers on hydrogeology ; 16)
    Includes bibliographical references and index.
    ISBN 978-0-415-54493-1 (hardcover : alk. paper) -- ISBN 978-0-203-85283-5 (e-book)
1. Groundwater flow. 2. Hydrogeology. I. Taniguchi, Makoto, 1930- II. Holman, I. P. (Ian Paul) III. Title. IV. Series.

  GB1197.7.G77 2010
  551.49--dc22

                                                                    2009044991

ISBN 13: 978-1-138-11259-9 (pbk)
ISBN 13: 978-0-415-54493-1 (hbk)

Table of contents

Preface

Groundwater is the world's largest, accessible store of freshwater. It is the primary source of drinking water to nearly half of the world's population; a vital source of irrigation water to contribute to global food security and helps to maintain the delivery of vital ecosystem services from rivers, wetlands and lakes.

Effective groundwater management requires an understanding of the effects of natural climate variability on groundwater levels, fluxes and quality, particularly in the context of increasing climate uncertainty associated with global climate change. Although the potential impacts of climate change on groundwater resources have long been recognised, the Inter-governmental Panel on Climate Change (IPCC) has noted in its Second, Third and Fourth Assessment Reports that there has been comparatively little research relating to groundwater.

In partial response to this, two important initiatives were started in the early part of this century—the UNESCO Groundwater Resources Assessment under the Pressures of Humanity and Climate Change (GRAPHIC) programme and the International Association of Hydrogeologists' (IAH) Commission on Groundwater and Climate Change.

This book focuses on integrating our knowledge of the relationships between climate change/variability and sea level rise and groundwater storage, recharge, discharge, and groundwater quality, based on case studies reported from around the world including Egypt (Barrocu & Dahab), Czech Republic (Novicky et al.), Japan (e.g. Hu et al.), Morocco (Ouysse et al.) and Indonesia (Abidin et al.). The knowledge from contemporary field investigations (Abidin et al.), water balance assessments (Hasegawa et al., Hu et al.), numerical simulations (Yoshimatsu et al.), satellite data (Ouysse et al.), paleohydrology (Teramoto et al., Iwatsuki et al., Hasegawa et al.), stable and radioactive isotope analyses (Kobayshi et al., Hasegawa et al., Machida et al., Yabusaki et al.) and monitoring studies (Taniguchi et al.) are reported. This book also examines the assessment of uncertainty in hydrological models that are driven by climate data (Novicky et al.).

Most of the papers were presented at the 36th IAH congress, which was held in Toyama, Japan on October 26th to November 1st, 2008. The papers were presented in Session S21 "Response of groundwater system from climate change, S22 "Impact of sea level rise on groundwater systems, S23 "Knowledge from paleo-hydrology, and Special Sessions 01" UNESCO-GRAPHIC" and 02 "Alluvial fans".

We acknowledge the organizing committee, executive committee, and scientific committee of the 36th IAH Congress, as well as the Japanese IAH sector. We wish to thank UNESCO-IHP and the Research Institute for Humanity and Nature (RIHN) for supporting the UNESCO-GRAPHIC session. We also acknowledge Yoko Horie (RIHN) and Janjaap Blom (CRC Press/Balkema–Taylor and Francis Group) for helping this publication.

Editors
Makoto Taniguchi
Research Institute for Humanity and Nature (RIHN)
and Chairman of the GRAPHIC programme

Ian Holman
Cranfield University and Co-Chairman of the IAH
Commission on Groundwater and Climate Change

REFEREES

The Editors are grateful to the following people for their assistance with the reviewing of papers submitted to this publication:

Jason Gurdak, *U.S. Geological Survey*
Ian Holman, *Cranfield University*
Shiho Yabusakai, *Rissho University*
Tetsuya Hiyama, *Nagoya University*
Yoichi Fukuda, *Kyoto University*
Tsutomu Yamanaka, *University of Tsukuba*
Isao Machida, *Geological Survey of Japan*
Masako Teramoto, *NIPPON KOEI Co., Ltd.*
Michael Valk, *IHP & HWRP*
Bret Bruce, *U.S. Geological Survey*
Teruki Iwatsuki, *Japan Atomic Energy Agency*
Jun Yasumoto, *Ryukyu University*
Yossi Yiechieli, *Geological Survey of Israel*
Makoto Taniguchi, *Research Institute for Humanity and Nature*
Satoshi Nakada, *Research Institute for Humanity and Nature*
Takeo Oonishi, *Research Institute for Humanity and Nature*
Takashi Kume, *Research Institute for Humanity and Nature*

About the editors

Dr. Makoto Taniguchi is a Professor of hydrology and a project leader of "Human impacts on urban subsurface environment" in RIHN. He is also a leader of UNESCO-GRAPHIC Project "Groundwater Resources Assessment under the Pressures of Humanity and Climate Change", and a vice president of the International Committee of Groundwater of IAHS/IUGG. He has published several books and many papers in international journals of hydrology, geophysics and environmental sciences.

Dr. Ian Holman is a Senior Lecturer and head of the Integrated Land and Water Group within the Natural Resources Department of Cranfield University, UK. As a hydrogeologist and climate impacts modeller, his principal research interest has been the holistic assessment of climate change impacts on hydrogeological and water resource systems. He led the UK's first regional integrated assessment of the impacts of climate and socio-economic change on agriculture, water resources, flooding and biodiversity, and presently serves as Co-Chairman of the International Association of Hydrogeologists' (IAH) Commission on Groundwater and Climate Change.

CHAPTER 1

Vulnerability of groundwater resources in different hydrogeological conditions to climate change

Oldrich Novicky & Ladislav Kasparek
T.G. Masaryk Water Research Institute, Czech Republic

Jan Uhlik
PROGEO s.r.o., Czech Republic

ABSTRACT: During recent years, T.G. Masaryk Water Research Institute in Prague has carried out a number of studies which were focused on the possible impacts of climate change on groundwater resources. For these studies, the Bilan water balance model was used to simulate the water cycle components (including groundwater recharge and base flow), both for conditions unaffected by climate change and also for those modified according to climate change scenarios. The initial studies have shown that groundwater resources in unfavourable hydrogeological settings (e.g. those in crystalline geological formations) are highly sensitive to climate change and can rapidly be exhausted. Subsequent applications of the Bilan model in combination with MODFLOW (modular three-dimensional finite-difference groundwater flow model developed by the United States Geological Survey) however showed that climate change could have dramatic consequences, particularly in basins with good hydrogeological settings (such as those in cretaceous geological formations), mainly with respect to groundwater depletions that will greatly affect the availability of water supply.

Keywords: Bilan, MODFLOW, groundwater, climate change

1 INTRODUCTION

For studies of possible impacts of climate change on water resources, the T.G. Masaryk Water Research Institute, p.r.i., uses the Bilan hydrological model, which was developed by staff of the Institute. The Bilan model is described in detail by Tallaksen and Lannen (2004). The executable version is available with example data sets on the CD that is attached to the textbook. The primary input data for the Bilan model include time series of monthly precipitation, temperature and relative air humidity. The lumped model simulates the water budget at three vertical levels: on the land surface, in the soil layer and in the groundwater aquifer. Three water balance algorithms are applied for winter conditions, snow melting and summer conditions. The surface water balance depends on actual evapotranspiration, which is determined from water availability and potential evapotranspiration calculated from meteorological conditions. For calculation of the potential evapotranspiration, empirical values that were derived in Gidrometeoizdat (1976) for different climate zones are used. Excess water (precipitation minus evapotranspiration) forms direct runoff or infiltrates to the deeper zone, where it is divided into interflow and groundwater recharge.

The outputs of the model include monthly series of water storage in the snow pack, soil and aquifer. Furthermore, surface runoff, interflow, and base flow (groundwater discharge) are calculated at the outlet of the catchment. The eight free parameters of the Bilan model are calibrated by minimising the differences between simulated and observed outflow from the basin.

In 1998, T.G. Masaryk Water Research Institute completed a study (Kašpárek, 1998) that used the Bilan model for the analysis of the possible impacts of climate change in the Czech part of Elbe River basin. Kašpárek (1998) divided the basin into 18 sub-basins and concluded that the least vulnerable sub-catchments are mountain basins in North Bohemia and in the Šumava Mountains. Basins that have a low precipitation rate and small retention capacity (both natural in groundwater bodies and artificial in reservoirs) would be most affected by climate warming.

The Bilan model was subsequently applied by Krátká and Kašpárek (2005) to data that were prepared for the period 1971–2002 by the Czech Hydrometeorological Institute for the basins of 50 water gauging stations from the whole territory of the Czech Republic. This study substantiated the above results in terms of the spatial distribution of the decrease in runoff, and concluded that decreases in minimum monthly flows were greater than decreases in mean monthly flows. For some basins and an unfavourable climate change scenario (assuming high CO_2 emissions and high climate sensitivity to CO_2 concentration), the minimum flow decreased by as much as 12% to 15% of the current flow. The percentage decrease in mean base flow was equal to or greater than the decrease in total runoff, which decreased from 6% to 43% depending on the climate change scenario and the physical conditions of the basin.

The results of a groundwater study by Kněžek and Krátká (2005) indicated that groundwater resources in unfavourable hydrogeological settings, such as those of a crystalline geological formation with shallow groundwater circulation, are highly sensitive to climate conditions or human influences. The groundwater resources in these systems can be rapidly exhausted under such climate change.

The attention in the subsequent studies was primarily aimed at examining possible impacts on groundwater resources in favourable hydrogeological settings, such as those in Cretaceous geological formations. This was initiated by the analyses of data from this type of basin which indicated that such groundwater resources could be also highly vulnerable. This can be illustrated for a selected study basin, whose groundwater storage did not drop below 94.8% of the mean value in the period 1976 to 1990, but decreased by 37% during the 2004 dry year (Kněžek and Krátká, 2005).

A combination of Bilan and MODFLOW (MODFLOW-2000, Harbaugh et al., 2000) was used to estimate the possible impacts of climate change on groundwater resources in the Metuje River basin (Novický et al., 2007), which is part of an important hydrogeological region (Cretaceous geological formation characterised by deep circulation of groundwater and high storage) in the Czech Republic. The results of a comparison of MODFLOW simulations using recharge series unaffected and affected by climate change showed that climate change could cause groundwater levels in the basin to decrease by as much as 10 metres. The base flow could drop to the levels of the existing groundwater abstractions in the basin (about 0.1 m^3 s^{-1}) and therefore the Metuje River would be dry in the periods when it is normally almost exclusively fed from groundwater storage.

A subsequent study was focused on the possible impacts of climate change in the Budějovická basin (Uhlík, 2008). Selected results of this study are described in this paper.

2   METHODS

2.1   *Climate change scenarios*

The climate change scenarios are based on the results of simulations by the HIRHAM (Christensen and van Meijgaard, 1992; Christensen et al., 1996) and RCAO (Döscher et al., 2002) regional climate models (RCM). The HIRHAM RCM uses the boundary conditions from HadCM3 (Gordon et al., 2000) GCM simulations and the RCM RCAO uses ECHAM4 (Roeckner et al., 1996) and OPYC (Ocean and isoPYCnal coordinates) GCM boundary conditions. These simulations use ocean temperatures for 2071–2100 from the GCM outputs using SRES emission scenarios (A2 as pessimistic and B2 as optimistic alternative) developed by Nakicenovic et al. (2000). Low (2°C) and high (4.5°C) climate sensitivity to CO_2 concentration (climate sensitivity refers to the equilibrium change in global mean surface temperature following a doubling of the atmospheric CO_2 concentration) were used in the B2 and A2 scenarios, respectively.

The spatial resolution of the regional models is about 50×50 km (the area of the Czech Republic is covered by about 50 points). The simulations were performed for the period 2071–2100 (time horizon of 2085) with a reference period of 1960–1990. The climate change scenarios were prepared in the form of monthly time series of absolute changes (delta factor) in air temperature and dew-point temperature and relative changes (percentage change factor) in precipitation.

For the simulation of water cycle components for current conditions, the Bilan model used observed weather data (for the study, period 1942–2007 was available), which are subsequently modified by the monthly change factors for the climate change scenarios for the simulation of water cycle components affected by climate change.

2.2   *Linked application of Bilan and MODFLOW model*

For the purpose of the groundwater studies, the Bilan model was coupled with MODFLOW. In this application, the spatially uniform groundwater recharge series simulated for current and climate change conditions by the Bilan model were used as an input to MODFLOW. The parameters of the MODFLOW model were calibrated by minimising deviations between observed and simulated groundwater levels and also between simulated base flow (from the Bilan model) and base flow derived from streamflow hydrographs. Steady state simulation by the MODFLOW model was subsequently used to simulate the spatial distribution of groundwater levels and base flow in the study basin. This approach is described in more detail in Novický et al. (2007).

3   STUDY BASIN AND DATA

The Budějovická basin is located (Fig. 1) in the southern part of the Czech Republic. The total area of the basin is 484 km^2; 240 km^2 of which is one of the most important hydrogeological formations of groundwater resources in Southern Bohemia. The groundwater in the basin has accumulated in the upper Cretaceous and Tertiary sediments, which are located on crystalline bedrock whose upper surface is at between 50 m and 380 m above mean sea level (a.m.s.l.). The elevation of the land surface ranges between 380 and 460 m a.m.s.l. Contour lines (isolines) of the bottom of the Cretaceous and Tertiary sediments are shown in Fig. 2 with the locations of groundwater observation sites and the main groundwater abstraction sites.

The mean annual air temperature is 7.9°C. The mean annual precipitation is 620 mm, of which 126.5 mm forms the total runoff (mean base flow is 35.4 mm). The groundwater storage is fed from recharge and partially from inflow from the surrounding crystalline rock formations.

Input data for the Bilan model (monthly series of basin precipitation, air temperature and relative air humidity) and flow series for calibration of the parameters of the model were available for the period 1942–2007. Observed groundwater levels in boreholes in the

Figure 1.   Location of the study area (Budějovická basin).

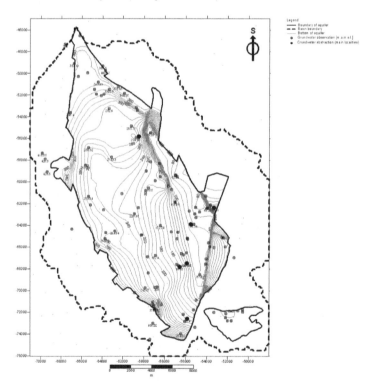

Figure 2.   Contour lines (isolines) of the bottom of the cretaceous and tertiary sediments in the Budějovická basin and the locations of groundwater observation and groundwater abstraction sites.

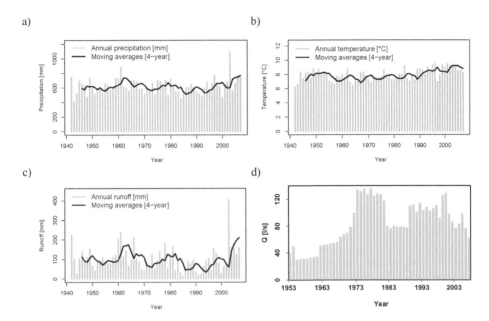

Figure 3. Results of hydrometeorological observations in České Budějovice (a—annual precipitation, b—annual air temperature, c—annual runoff from the Budějovická basin, d—annual groundwater abstractions (l/s) from Budějovická basin).

basin since 1974 were available for calibration of the MODFLOW model. Fig. 3 illustrates selected hydrometeorological observations and data on groundwater abstractions.

4 RESULTS

4.1 *Analysis of observed data*

During the last 20 years, mean annual air temperature increased by about 1°C (Fig. 3). Because of an increase in mean annual precipitation, the increase in mean annual air temperature was not reflected in a decrease in annual runoff. High annual runoff in 2002 is related to extreme flood conditions during the year. The increased temperature in combination with mean or low precipitation after 2000 (except for 2002) was probably reflected in a drop of groundwater levels. These results are consistent with results for other basins in the Czech Republic (Uhlík, 2006).

4.2 *Bilan model simulation*

The capability of the Bilan model to simulate the water cycle components in the basin is illustrated in Fig. 4, which shows the flow duration curves that were derived from the observed and simulated monthly flow series. The fit between the curves in the low flows is very good while high flows are moderately underestimated by the simulation. This is reflected in long-term mean annual runoff, which is 126.5 mm yr^{-1} when derived from the observed data and 112.8 mm yr^{-1} calculated from the simulation.

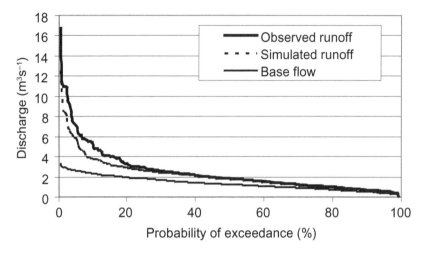

Figure 4.   Flow duration curves derived from monthly observed and simulated data and base flow simulated by the Bilan model.

All of the climate change scenarios predict large increases in the mean annual air temperature (of between 3.7°C and 5.2°C) and moderate decreases or increases in mean annual precipitation depending on the scenario (Table 1). The seasonal distribution of precipitation could significantly change, which would be unfavourable because a decrease is predicted for the summer season when drought periods frequently occur. The changes in climate variables would be reflected in substantial deterioration of hydrological conditions according to the results of the Bilan model simulations. The annual average total runoff would drop to between 28 and 58 mm/year, base flow to 5 to 10 mm/year and groundwater recharge (an important input to the MODFLOW model) would also dramatically drop from the current annual average value of 35 mm/year to 5 to 10 mm/year (14% to 28% of the current value) (Table 1).

4.3   *MODFLOW model simulation*

3D groundwater flow was simulated by dividing the Budějovická basin into four model layers in MODFLOW, with differing coefficients of hydraulic conductivity. The elevation of the bottom and top surfaces of the layers are shown in Fig. 5. Fig. 6 shows the distribution of the infiltrated precipitation (l/s) between the individual layers. Under current conditions, the average infiltration into the aquifer is 523.6 l/s, of which 367.4 l/s forms shallow water circulation in the upper part of the aquifer and 156.2 l/s percolates into the deeper aquifer parts. With climate change, the infiltration into layer 1 would drop to between 73 and 145 l/s depending on the scenario. From 1990 the total groundwater abstraction decreased from 100 l/s to current 70 l/s. Budweiser bier abstracts constantly 20 l/s.

The original goal of the study was to simulate the impacts of the existing groundwater abstractions in combination with the scenarios of climate change. But predicted infiltration under climate change (73 l/s–145 l/s) will be almost equal to the current groundwater abstraction. Therefore, the original study goals were changed and the groundwater flow was simulated by MODFLOW alternatively for conditions involving groundwater abstraction

Table 1.   Average annual climate parameters for current and future conditions and main hydrological results of the Bilan model simulations.

Model	SRES scenario	Precipitation [mm/year]	Temperature [°C]	Potential evapotranspiration [mm/year]	Runoff [mm/year]	Base flow [mm/year]	Recharge [mm/year]
RCAO	A2*	582.4	13.09	931	28	5.63	4.91
RCAO	B2*	634.5	11.66	815	48.3	8.5	7.82
HIRHAM	A2*	612.4	13.05	905	36.4	5.44	6.33
HIRHAM	B2*	658.3	11.63	803	57.66	10.41	9.77
Current conditions (1942–2007)		619.9	7.89	673	100.9	35.44	35.16

* Current weather (1942–2007) modified according to change factors for 2071–2100.

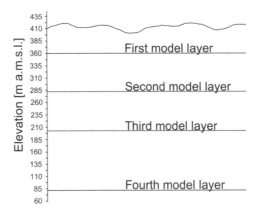

Figure 5.   Elevation of the bottom and top surfaces of the MODFLOW model layers.

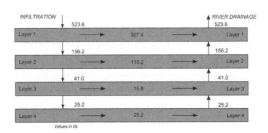

Figure 6.   Distribution of the infiltrated water (l/s) between the individual aquifer layers under the current climate (1942–2007).

(but not climate change) and those reflecting climate change scenarios (but not groundwater abstraction).

Fig. 7a illustrates the results of the MODFLOW simulations for conditions of current groundwater abstractions, and indicates that groundwater level in the well screen depth dropped by as much as 10 m in the vicinity of the main groundwater-abstraction sites. These sites are located close to the centre of the basin and affect the groundwater conditions mainly in the model layer 3. The maximum decreases in groundwater levels caused by

a)

b)

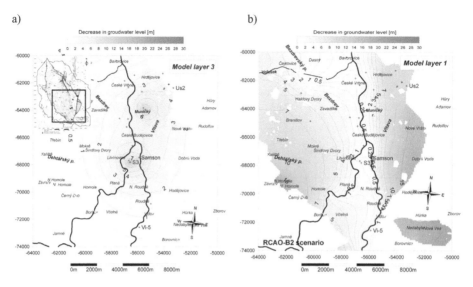

Figure 7.   Decrease in groundwater levels caused by (a) groundwater abstraction under the current climate and (b) the B2 climate change scenario from the RCAO model (without abstraction).

climate change occur in the areas located close to the basin boundary. These decreases, which reach 15 and 25 m depending on the climate-change scenarios, affect mainly the groundwater conditions in model layer 1 (Fig. 7b).

5   DISCUSSION AND CONCLUSIONS

The results of the study are consistent with findings from previous studies that groundwater resources in good hydrogeological settings are highly vulnerable to groundwater level declines, particularly when they are exposed to a combination of impacts from abstraction and climate change. Under likely climate-change scenarios for the Budějovická basin, it will probably not be possible to meet the existing requirements for groundwater abstraction and therefore other resources will have to be used to meet the water-supply demand. Thus, conjunctive use of groundwater and surface-water resources will be necessary in the future for successful water resource management because surface water resources will be also limited. In addition, water demands are likely to increase, particularly for some purposes such as irrigation of agricultural lands. The conclusions derived from the results of this and future studies and proposed measures must be incorporated into strategies aimed at reducing the impacts of climate change, particularly for water management planning. In the member countries of the European Union, these results can suitably be applied in the implementation of the Water Framework Directive and in meeting its requirements for the development of Programmes of Measures focused on protection of water resources and their sustainable use.

The results of the simulation of the impacts of climate change on groundwater resources are naturally affected by a number of uncertainties, which originate from inaccuracies in input hydrological and meteorological data, the time step used for the simulation, uncertainties in hydrological model simulations, and particularly those stemming from the climate model simulations. The main drawback of the existing climate models is their insufficient capability to simulate the spatial distribution of precipitation. Data on the simulated total runoff in Table 1 suggest that the key uncertainties originate from different assumptions concerning CO_2 emissions.

ACKNOWLEDGEMENT

This paper was prepared from the results of research projects sponsored by the Ministry of the Environment of the Czech Republic, particularly the project on Research and protection of hydrosphere—research of relationships and processes in water component of the environment focused on impacts of human pressures, the sustainable use and protection of the hydrosphere and legislative tools (project No. MZP0002071101) and the project on Refining of current estimates of impacts of climate change in sectors of water management, agriculture and forestry and proposals of adaptation measures (project No. SP/1a6/108/07).

REFERENCES

Christensen, J.H., van Meijgaard, E. (1992) On the construction of a regional climate model, Tech. Rep., 92–14, DMI, Copenhagen, 22 p.
Christensen, J.H., Christensen, O.B., Polez, P., van Meijgaard, E., Botzet, M. (1996) The HIRHAM4 regional atmospheric Climate model, Sci. Rep., 96-4, DMI, Copenhagen, 51 p.
Döscher, R., Willén, U., Jones, C., Rutgersson, A., Meier, H.E.M., Hansson, U., Graham L.P. (2002) The development of the coupled regional ocean-atmosphere model RCAO, Boreal Env. Res. Vol. 7: 183–192.
Gidrometeoizdat (1976) Rekomendatsii po roschotu ispareniia s poverhnosti suchi. Gidrometeoizdat St Peterburg.
Gordon, C., Cooper, C., Senior, C.A., Banks, H., Gregory, J.M., Johns, T.C., Mitchell, J.F.B., Wood, R.A. (2000) The simulation of SST, sea ice extents and ocean heat transports in a version of the Hadley Centre coupled model without flux adjustments, Climate Dynamics, Vol. 16: 147–168.
Harbaugh, A.W., Banta, E.R., Hill, M.C., McDonald, M.G. (2000) MODFLOW-2000, the U.S. Geological Survey modular ground-water model—User guide to modularization concepts and the Ground-Water Flow Process: U.S. Geological Survey Open-File Report 00-92, 121 p.
Kašpárek, L. (1998) Regional study on impact of climate change on hydrological conditions in the Czech Republic. Prague, VÚV T.G.M., 1998, ISBN 80-85900-22-X, 69 p.
Kněžek, M., Krátká, M. (2005) Quantification of groundwater regime during extreme hydrological situations. In: Hydrological days 2005, Slovak Hydrometeorological Institute and Slovak National Committee for hydrology, Bratislava, ISBN 80-88907-53-5 [in Czech].
Krátká, M., Kašpárek, L. (2005) Regional impacts of climate change on water regime in the CR. In: Hydrological days 2005, Slovak Hydrometeorological Institute and Slovak National Committee for hydrology, Bratislava, ISBN 80-88907-53-5 [in Czech].
Nakicenovic, N., Alcamo, J., Davis, G., de Vries, B., Fenhann, J., Gaffin, S., Gregory, K., Grübler, A., Jung, T.I., Kram, T., Lebre La Rovere, E., Michaelis, L., Mori, S., Morita, T., Pepper, W., Pitcher, H., Price, L., Riahi, K., Roehrl, A., Rogner, H.H., Sankovski, A., Schlesinger, M., Shukla, P., Smith, S., Swart, R., van Rooijen, S., Victor, N., Dadi, Z. (2000) IPCC Special Report on Emissions Scenarios, Cambridge University Press, Cambridge, United Kingdom and New York, NY.
Novický, O., Kašpárek, L., Uhlík, J. (2007) Possible impacts of climate change on groundwater resources and groundwater flow in well developed water bearing aquifers. In: Proceedings from The third international conference on climate and water. Helsinki, Finland, 3–6 September 2007, ISBN 978-952-11-2790-8.

Roeckner, E., Oberhuber, J.M., Bacher, A., Christoph, M., Kirchner, I. (1996) ENSO variability and atmospheric response in a global coupled atmosphere-ocean GCM. Clim. Dyn. Vol. 12: 737–754.

Tallaksen, L., Lannen, H. (editors) 2004 Hydrological drought—processes and estimation methods for streamflow and groundwater. Developments in water science, 48, Elsevier B.V., Amsterdam.

Uhlík, J. (2006) Analysis of impacts of climate change in central part of Intra-Sudeten basin. PROGEO and T.G. Masaryk Water Research Institute, Prague. [in Czech].

Uhlík, J. (2008) Possible impacts of climate change on groundwater resources in Budějovická basin. PROGEO and T.G. Masaryk Water Research Institute, Prague. [in Czech].

CHAPTER 2

Changing climate and saltwater intrusion in the Nile Delta, Egypt

Giovanni Barrocu
Department of Land Engineering, University of Cagliari, Italy

Kamal Dahab
Geology Department, Faculty of Science, Menoufia University, Shebin El-Kom, Egypt

ABSTRACT: The Nile delta aquifer, consisting of sediments deposited by the most important river flowing into the Mediterranean, provides the main water supply for nine Governorates in the area. Groundwater is endangered by salt water encroachment due to lateral intrusion of present sea water and upconing of connate salt water trapped in paleodeltaic sediments. The interface between sea and groundwater with different salinity is very fragile, as a result of alternate sea level fluctuations due to climate changes over geological time. The natural water cycle and river sedimentation in the delta have been strongly affected by human actions, namely the construction of the Aswan High Dam, and pollution. Therefore, integrated surface water and groundwater resources of different salinity should be rationally managed in terms of quantity and quality for different uses to face the ever-growing demand of the increasing population in the Nile Delta, who make a living from intensive agriculture.

Keywords: Climate change, salt water, Nile Delta

1 INTRODUCTION

The Nile Delta ($\sim$22,000 km^2) constitutes a large leaky aquifer system, representing the most important groundwater resource for nine Governorates of Egypt. Its apex is located at about 20 km north of Cairo and 180 km from its base, stretching around 220 km along the Mediterranean seashore.

The closure of the Aswan High Dam disrupted the natural surface and groundwater flow and sediment discharge cycle, so that the northern part of the delta is endangered by erosion, saltwater intrusion, and pollution. These processes are inducing loss of both land and coastal lagoons, and a marked decline in agricultural productivity at a time when the population is expanding exponentially. In order to assess the impacts of the natural processes and human actions on groundwater and the extension and origin of salt water intrusion in the aquifers of the Nile Delta, a network of observation wells, irrigation and domestic pumping stations was developed and data on salinity and hydrochemistry collected by one of the co-authors in a study area between Longitude 30°30′ and 31°30′ E, and Latitude 31°00′ and 31°30′ N were reassessed (Fig. 1).

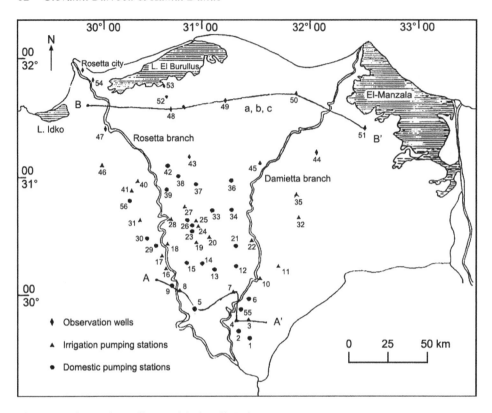

Figure 1.   Observation well network in the Nile Delta.

2   GEOLOGICAL SETTING

The present day delta is the result of different sedimentation phases from the Late Pliocene to Present associated with alternating sea regressions and ingressions due to climate changes and tectonic movements, with down faulting development or rejuvenation of older faults, affecting the Mesozoic bedrock and overlaying sediments deposited by the Nile. Down-slope mass movements, sediment slumping and basin bottom subsidence gave space to a highly developed deltaic front underlined by the prodelta formed by the deposits of the old Nile, with its tributaries from the Eastern and western deserts, and some local tributaries (Sestini, 1989).

The Plio-Quaternary aquifer system of the Nile Delta overlays the thick sequence of Pliocene, Messinian (El Wastani, Kafr El-Sheikh, Abu Madi) and older Tertiary and Cretaceous strata (Fig. 2). Stratigraphically, the system can be easily distinguished according to lithological characteristics into two rock units, from top to bottom:

a. *The Bilqas Formation*, belonging to the Holocene, is mainly composed of silt and sandy or clayey silt with few sand layers in the lower part. Also some calcareous inclusions are detectable at different depths.

   In the Nile delta coastal area, sediments are fine and attain thicknesses of up to about 50 m, so that they act as an aquiclude. Southwards, they become gradually thinner, their

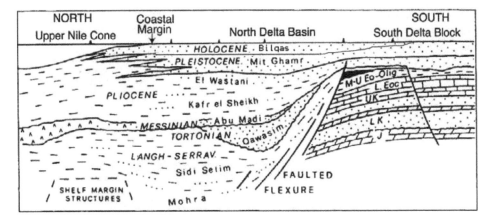

Figure 2. Cross-section from S-to-N showing Quaternary deposits above thick Tertiary and Mesozoic sequences (modified after Sestini, 1989).

thickness ranging between 20–25 m in the southern and middle parts, whereas their grain size becomes coarser, so that they act as an aquitard.

b. *The Mit Ghamr Formation,* assumed to be of Late Pliocene—Early Pleistocene age, consists of sands and gravels interbedded by thin clay layers and lenses. Locally, the grain size of the sediments change so that they predominantly consist of clay and silt with sand intercalations. The funnel-shaped depositional basin of the formation covers most of the study area and its fringes, its long stem extending southwards into the Nile valley. Clays are more frequent in the north, whereas coarser sediments dominate in the southern part and are occasional in the eastern and western sides. The series of alternating clays and sands with gravel, present especially in the northern and north-eastern parts, reflects the progradation cyclicity sequence of the Delta. The formation, over 900 m thick in the northern part of the Nile Delta, 500–600 in the middle part, and 100–400 m in the southern part, constitutes the main aquifer of the region.

Sedimentological research carried out on a number of core samples from lagoon/marsh, delta front and prodelta facies in the northeast sector of the region clearly shows that the deltaic deposits mainly consist of detritus and weathering products from the Trap Series Basalts covering the 75% of the Ethiopian Highlands (Siegel et al., 1995).

In Fig. 3, the Nile Delta development phases due to sea level variations caused by climate changes from late Pliocene to the beginning of the Holocene are outlined over the image of the present delta.

During the very strong pluvial periods at the end of the Pliocene, the Nile deposited enormous quantities of sediment in the pre-existing sea gulf. During the Early and Middle Pleistocene, the main bulk of the Nile Delta was built under fluviomarine conditions, and nearly all porosities were saturated with salt water. The thickness of fresh water-saturated sediments was very limited and existed only in the earliest delta lobes to the southeast and southwest. By the end of the Middle Pleistocene, the Mediterranean Sea had reached a level similar to the present one or even lower. Such progressive sea level lowering determined the leaching of the old Pleistocene aquifers through the dilution of the inherited salt water by the infiltrated rain water discharging into the effluent channel (Dahab, 1994).

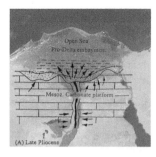
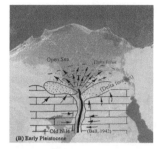
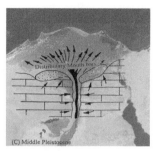
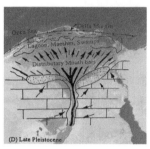
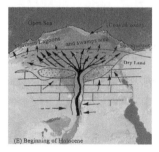

Figure 3.   The Nile Delta development due to climate changes from late Pliocene to Holocene outlined over present delta satellite images. A) Late Pliocene: 2.6-1.8 My. B) Early Pleistocene: 1.8-0.78 My. C) Middle Pleistocene: 0.78-0.12 My. D) Late Pleistocene: 0.12-0.0017 My. E) Beginning of Holocene: 0.0017 My.

The most drastic changes in the Nile Delta aquifer system took place during the late Pleistocene (nearly 180,000 years Before Present (B.P.)), owing to severe changes in sea level and climatic conditions. At least eight stages are of certain importance for groundwater quality evolution:

i.   Interglacial episodes separating the Middle and Late Pleistocene (Mindel-Riss inter-glacial) caused a sea level rise above its present day level by about 8 m (Fairbridge, 1961). This caused coastline regression on the subaerial delta plain with a corre-sponding aggradation in the distributaries and on the low lands. Deterioration of groundwater occurred but, on the other hand, the channels in the Nile valley and the southern parts of the Nile Delta became influent and recharged the adjacent older sediments with fresh water, which leached out their inherited salinity.

ii. Geological and archaeological investigations indicate that the Nile Delta margin has been affected by submergence due to eustatic sea-level rise (~1 mm/year) since the late Holocene, and land subsidence at variable rates, confirmed by the submergence and burial of the base of most archaeological sites along the delta (Sestini, 1989; Stanley and Warne, 1998; Stanley, 2005). The interval between 180,000 and 125,000 yr B.P. was the time of the Riss glacial stage, associated with a notable drop in sea level, below the present one. This favoured the degradation of the main river courses and the effluent channels caused another cycle of leaching in the Early and Middle Pleistocene aquifers in the southern parts.

iii. In the time interval between 125,000 and 110,000 yr B.P., there was a maximum transgression, when the sea level rose up to 9–15 m above the present one, so that the transgressive sea spread over most central low lands. The pre-existing incised channels shortened but were also greatly aggraded, so that they caused the most effective recharging and leaching cycle of the older aquifers. Both groundwater levels and the thickness of the fresh water saturated zone reached their maximum.

iv. The time interval between the last Pleistocene maximum transgression and 30,000 yr B.P. was characterized by continuous sea level fluctuations around the present sea level. Much of the latest Pleistocene deltaic deposits were accumulated under fluviomarine conditions, their porosities being mainly saturated with salt to brackish water. The older aquifers were still existing under continuous leaching and the inherited salinity was progressively decreasing. Alternating degradational and aggradational phases, due to climatic and hydrological condition variation, caused the deposition of several interbedded clay layers, which generally increase in thickness northward. Accordingly, the upper part of the Quaternary aquifer system was split into several sub-aquifers with different piezometric heads.

v. From 30,000 to 13,000 yr B.P., the sea level sharply dropped by more than 50 m (maximum about −130 m below current sea level), and the coastline retreated northward nearly to the outer limit of the continental shelf. All distributaries deepened their thalweg and behaved as effluent channels. Accordingly, Quaternary aquifers were washed out and discharged almost completely their salt water (Dahab, 1994).

vi. From 13,000 to 6,000 yr B.P., the last sea transgression started, at first rapidly and then at a rate from moderate to low. The lowest ultimate base level was certainly accompanied by aggradation in the delta tributaries, and the southern parts of the Nile Delta aquifer system were recharged with fresh ground water.

vii. During the last 6,000 years, slight fluctuations have been recorded. From 4,000 to about 1,600 yr B.P. the average sea level was 2 m lower than today, so that in extensive tracts of the northern low lands to the west of Lake El-Burullus (Fig. 1) cultivation flourished and population numbers increased. Since 1600 yr B.P., the sea level has been continuously rising, but the sediment load of the Nile could counteract sea ingression so that the delta could continue growing. The river flow was enough also to block water and soil salinization processes endangering the northern parts of the delta.

viii. After closing the High Dam in 1968, floods and their associated suspended load were totally prevented. This led to severe coastal erosion, and coastal aquifer deterioration due to salt water intrusion. Following the hydrological and geomorphological variations produced by the High Dam, the Mediterranean Sea coastline is advancing inland. This process is due to the combination of global sea level rise at a rate of 1–5 mm/year (Fairbridge, 1961), and continuous subsidence, at about 1 mm/year, of the unstable northern part of the Nile Delta, where Nile sediments have been drastically reduced. The coastal sectors most endangered by erosion are the three narrow beach barriers separating the Mediterranean Sea from the Lakes Idko, El-Burullus and El-Manzala (Fig. 1).

Consequently, the Mediterranean coastline will be shifted to the southern boundaries of the present coastal lakes, all the reclaimed lands will be waterlogged, and suffer from severe salinization. The reclaimed Barari belt, to the south-east of the lake El-Burullus, will revert into inundated marshy lands. Moreover, the Quaternary sand aquifer in the central parts of the Nile Delta along the expected new coastline will be affected by salt water encroachment and fresh groundwater deterioration.

River hydrology, eustatism, subsidence and anthropogenic activities are the main factors controlling the nature of the groundwater reservoir in the Nile Delta. Among all the above mentioned factors, global glacio-eustatic changes were the most effective, especially during the Quaternary, as pointed out by several authors (Fairbridge, 1961; Shackleton and Opdyke, 1976; Dahab, 1994).

The potential impact of a sea level rise of 0.5–1 m on the Nile Delta coastal area is clearly shown by the UNEP satellite images (Otto Simonett, UNEP/GRID-Arendal, 2002).

3   HYDROGEOLOGY

The Bilqas formation includes local aquifers consisting of sand and silt lenses, generally of uncertain geometry, where groundwater is obtained by shallow wells (of 15–40 m depth). In the southern and middle parts, silt and clay levels interbedded in the formation act as aquitards with an estimated leakage factor ranging from 4,000 to 999 m (Dahab, 1994). The piezometric surface is 3.5–1.5 m deep in the top sandy clay aquitards of the southern and middle parts, and 0.25–0.35 m above the ground surface in the northern part (north of Kafer El-Shiekh, Fig. 4), so that this may cause soil salinization and water ponding. Water levels are recharged to a great extent by irrigation, so that they reach their maximum values near the irrigation canals during the irrigation periods. Locally, these aquifers are likely recharged by surface water infiltration ponds of waste waters discharged by villages lacking sanitary drainage networks. In most Nile Delta villages waste water ponding is considered the main cause of groundwater contamination (El-Menayar, 1999). In the lowest aquifer layers, consisting of old lagoonal deposits, water levels measured at different localities were below the present sea level and they generally decrease seawards, down to the values of -6.5 m and -2.18 m above sea level (m asl). These water levels do not change over time, indicating that this groundwater is connate and isolated from the Nile River and present day sea.

The main delta aquifer system is represented by the Mit Ghamr formation which, in the northern part, consists of three main sub-aquifers separated by large clay intercalations. The uppermost sub-aquifer, with a water table 1–20 m deep, is supplied by fresh surface water seeping from irrigation canals and Nile branches, and by a south-north fresh groundwater flux. Groundwater in the middle sub-aquifer, with heads ranging from 20 to nearly 100 m below sea level, consists of old connate water entrapped in the early Holocene (Dahab, 1994). Groundwater in the deep sub-aquifer, formed in Late Pleistocene, is recharged by a south-north fresh seepage flux.

The estimated transmissivity of the Plio-Pleistocene aquifer ranges from 3,000 to 9,000 m^2/day, the hydraulic conductivity from 60 to 70 m/day, and the storage coefficient from 0.0005 to 0.0009 (Dahab, 1994). Groundwater flow, deduced from the piezometric map, is generally directed from the south to the north, towards the sea and coastal lakes. Groundwater levels range from $+9.8$ to $+0.5$ m asl with an average hydraulic gradient of 4.5 cm/km, estimated in the northern part of the Nile Delta. The actual velocity of groundwater flow ranges between 6.7 and 5.9 cm/day. Along the channel of the Rosetta branch, the actual groundwater velocity reaches an average value of 21.42 cm/day.

Groundwater levels in the Pleistocene aquifer generally decrease northward, ranging from 5 to 2.8 m asl. Moreover, in the piezometers located in the northern margin groundwater overflows under artesian conditions. Levels fluctuate over time, as recorded in 1992 and 1996, depending on the Nile flow and affecting the interface size and shape.

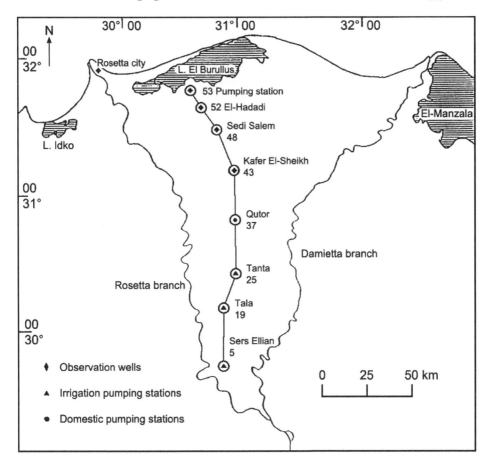

Figure 4.   Sketch map of the study area with the location and number of the sampling wells.

3.1 *Hydrogeological evolution of the Nile Delta after the Aswan High Dam construction*

The study of the recent surface water hydrology of the Nile Delta and its evolution indicated that:

1. After the High Dam construction, the annual surface water volumes entering the Nile Delta decreased from a range of $89.45 - 70.80 \times 10^9$ m³/year to a range of $40.73 - 35.18 \times 10^9$ m³/year. About 40% of these volumes, mostly suitable for irrigation, are discharged into the sea and coastal lakes.
2. After the High Dam construction, the average surface water levels of the Rosetta branch decreased from $+13$ to $+10.10$ m asl at the Delta barrage and from $+3.5$ to $+2.45$ m asl at the Kafer El-Ziat bridge.
3. Before the High Dam construction, the average annual surface water flow entering the Rosetta branch ranged between 30.56×10^9 and 31.66×10^9 m³/year. This represented 50% of the total inflow in the Nile Delta area. A flow of $27.02 - 26.26 \times 10^9$ m³/year (85%–75%) was discharged into the sea. After the High Dam construction, the flow

   ranges between 3.62 and 4.38×10^9 m^3/year, and a flow of $2.75 - 1.48 \times 10^9$ m^3/year (62%) is discharged into the sea.

4. Before the High dam construction, the Rosetta branch acted as a groundwater effluent stream in the dry season and as a groundwater influent stream in the flood season. After the High Dam, the branch acts as groundwater effluent stream throughout the year, particularly in its southern and middle parts. Considering the hydrologic equilibrium equation, the annual groundwater flow drained by the branch was estimated at 395×10^6 m^3/year.

5. Average surface water levels at the Damietta branch before the High Dam construction ranged from +13.78 m asl at the first location and +4.56 m asl at the second location. After the High Dam, the average levels of +11.23 m asl and +4.56 m asl were respectively measured at the same locations.

6. Surface water inflow into the Damietta branch ranged from 20.0 to 26.0×10^9 m^3/year before the High Dam, and afterwards it decreased to the range of 91.59 to 82.89×10^9 m^3/year.

7. At present, the Damietta branch contributes to recharge to the delta aquifer particularly in its southern and middle parts, and drains it downstream of the Zifta barrage.

8. Irrigation is considered to be the main source of groundwater recharge. The annual surface water seepage into the aquifer is estimated at 4.75×10^9 m^3/year.

9. In the northern part, drainage water has a high salinity, due to the interaction with salty lacustrine soils, so that it is unsuitable for irrigation without special treatment. The salinity of the water drained in the southern and middle parts is lower and within the limits required for irrigation.

10. The aquifer outflow was lower than its safe yield in the period from 1968 to 1978, and it was higher afterwards. This may be due to the decrease of surface water recharge and also to an increase in groundwater abstraction, so that the interface between sea water and surface and groundwater of different salinity has been strongly affected.

11. At present, the total fresh water inflow into the Nile Delta aquifer exceeds its total outflow of 2.5×10^9 m^3/year by an estimated 600.85×10^6 m^3/year. This groundwater flow is lost laterally towards the western side.

4   HYDROCHEMISTRY

4.1   *Methodology*

Eight groundwater samples were collected in 1993 from the deep wells (≥ 100 m) of the domestic pumping stations (Nos. 37, 43 and 52) and irrigation planting stations (Nos. 5, 19, 25, 48 and 53) located along a N-S section of the Nile Delta (Fig. 4). Wells, generally equipped with functioning pumps, were previously purged by bailing and pumping to remove suspended materials from the well and the immediate area outside the well screen. This procedure was continued until the pumped water was visually free of suspended materials or sediments. Groundwater level, temperature, pH and electrical conductivity were measured in the field. Water samples were also collected in the river Nile and sea (Dahab, 1994). Data were reassessed for the present study and compared to those previously obtained for the same wells in different periods by other authors (Farid, 1980; Saleh, 1981; Serage El-Din, 1989) (Fig. 5).

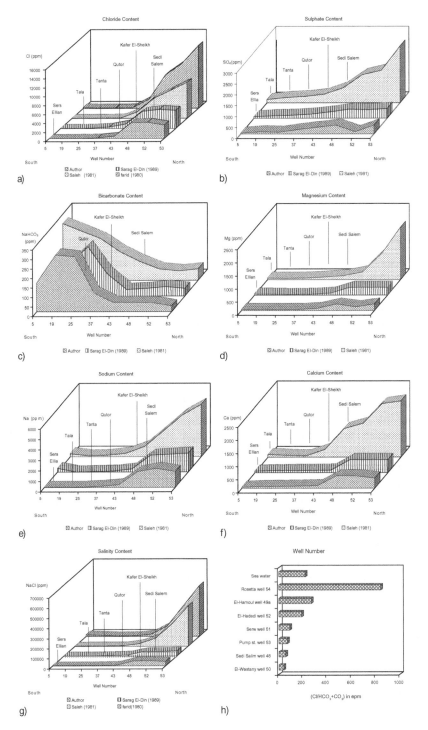

Figure 5.   Element contents and salinity distribution from the south to the north and comparison between the present and previous results along the hydrochemical south—north section in Fig. 4.

Samples were analyzed for major and trace elements in the central laboratories of TU-Berlin University and Menoufia University. Cations (Na, K, Mg, Ca, Br, B, Sr, Fe, Mn and Li) were determined using Inductively Coupled Plasma (I.C.P). Anions (Cl. SO_4, HCO_3) were determined by chromatography. Data were processed using standard methods for assessing groundwater quality and salt water intrusion.

Samples were also analyzed for stable isotopes (Oxygen and Deuterium) by mass-spectrometry at the laboratory of the Environmental Physics Institute of Heidelberg University, Germany. These data were used for identifying possible sources of salt water intrusion.

The hydrochemical data from the 54 wells of the observation network (Fig. 1) were processed in comparison with sea and Nile waters using Schoeller graph and Piper diagram (Schoeller, 1962; Piper, 1953) to identify the ion dominance and hydrochemical facies.

4.2   Results

The assessment of the hydrochemical data available for the aquifer system of the Plio-Pleistocene Mit Ghamr formation indicates that:

1. Groundwater Chloride content ranges from 35 to 310 ppm in the wells at the southern and middle parts of the study area, and from 2,261 to 4,194 ppm in the wells at the northern part (Fig. 5a).
2. Sulphate content ranges from 20 to 174 ppm in the southern and middle parts and from 200 to 590 ppm in the northern part (Fig. 5b).
3. Bicarbonate varies from 103 to 268 ppm in the southern part, and between 23 and 30 ppm in the northern part. The decrease in carbonates species is attributed to paleosea water intrusion and high mature stage of metasomatism (Fig. 5c).
4. Magnesium ranges from less than 20 to 40 ppm in the southern part, and from 105 to 265 ppm in the northern part. The increase in magnesium concentration in the northern part indicates sea water intrusion (Fig. 5d).
5. Sodium ranges from 23 to 52 ppm in the southern and middle parts and from 1,192 to 2,016 ppm in the northern part. High sodium concentration in this part reflects a marine origin (Fig. 5e).
6. Calcium ranges from 50 to 90 ppm in the southern and middle parts, indicating meteoric origin water. In the northern part it varies from 183 to 550 ppm in deep wells (Fig. 5f).
7. Different water zones with different salinity have been recognized (Fig. 5g). Samples collected from the domestic well No. 49 show salinity contents varying from 275 to 370 ppm. The water samples collected from the irrigation pump No. 8 show salinity from 271 to 432 ppm. In the northern part, salinity is higher, ranging between 4,100 and 7,000 ppm in a 105 m deep well (No. 49) and from 77,000 to 56,789 ppm in a shallow well 40 m deep (No. 50).

   In the wells of Tala (No. 19) and Tanta (No. 25) (Figs. 4 and 5g) the maximum salinity value reaches 650 ppm, and salinity measurements in test wells 350 m deep, drilled in their surroundings, showed a contour line of 1,000 ppm. At the El-Hamoul observation well (100 m deep), located 20 km north of the well of Kafer El-Shiekh (No. 43), a maximum salinity of 9,100 ppm was measured. A value of 35,000 ppm was found at the depth of 80 m in Rosetta city, 2 km from the sea. These high values are due to connate water entrapped under past drier climatic conditions.

8. The groundwater samples collected from the deep wells of the northern part (Nos. 52, 51, 53, 48 and 50) show low values of the $Cl/(HCO_3 + CO_3)$ coefficient compared to present-day seawater (230 epm) with the exception of the El-Hamoul well (No. 49a). This indicates that this water type was formed under drier conditions, and has been completely separated from the present day seawater for a considerable time (Figs. 1 and 5h).

9. Salinity changes with depth, as shown in the cross sections A–A' and B–B' in the southern and northern part of the Nile Delta (Figs. 1 and 6). Differences in salinity in the southern part, between wells No. 5 and 8, are likely due to local horizontal and vertical variations in the interface zone, as these wells are not along a straight line. High values in the top layer of section B–B' indicate hypersaline fossil waters in isolated sandy-loam lenses interbedded in clay layers of the Holocene Bilqas formation (Fig. 6). Such waters, characterized by salinity values ranging from 27,706 to 77,681 ppm, were trapped under past drier conditions and are completely separated from present saltwater and fresh water bodies.

   The uppermost sub-aquifer of the Mit Ghamr formation is strongly affected by sea water intrusion from coastal lagoons and the Mediterranean Sea. Its groundwater, from fairly fresh to brackish, with salinity ranging from 1,500 to 3,000 ppm is intruded by salt water from the sea and coastal lagoons of Idko, El-Burullus and El-Manzala (Dahab, 1994). The groundwater of the deep sub-aquifer, formed in the Late Pleistocene, is brackish, with a salinity ranging from 3,000 to 7,700 ppm.

   Groundwater is recharged by a south-north fresh seepage flux from the river Nile and its irrigation canals and is affected by present sea water intrusion. Fresh groundwater floats on an interface of brackish groundwater with a salinity of 1,000 ppm. The interface depth ranges from 200 m in the area of Tanta (well No. 25) to 100 m at Qutor (well No. 25), in the middle part of the Delta. Head and salinity fluctuations were observed over time (Serage El-Din, 1989).

10. Hydrochemical analysis shows that groundwater samples of the Plio-Pleistocene aquifer collected in the northern part from deep observation wells (105 m) have a typical Na-Cl facies with high Mg and Ca (Figs. 1 and 7a). Groundwater sampled from the transitional zone in the delta between the wells of Kutor (No. 37) and Kafer El-Sheikh (No. 43) are ascribed to a Na, Ca, Mg-Cl facies with high HCO_3 (Figs. 3 and 7b). Samples collected from the wells of the irrigation pumping stations (No. 7, 10, 11, 19, 20, 22, 24, 25) at the depth of 90 m have a typical Na, Ca, Mg-HCO_3 Cl facies (Figs. 1 and 7c). Samples from the wells of the domestic pumping stations 40–67 m deep (No. 1, 2, 5, 6, 9, 12, 13, 21, 29, 30, 33, 34, 55, 56) are characterized by a Na, Mg, Ca-Cl HCO_3 facies with high SO_4 (Fig. 1 and 7d).

11. Alkaline earth (Ca and Mg)—bicarbonate, sulphate water facies are dominant in the southern and middle parts, represented by areas A and B, respectively in Fig. 7e and f. This reflects leaching and dissolution by fresh Nile water during the Early and Middle Pleistocene. The Ca, Mg, Na-SO_4 + Chloride water facies, dominant in the northern part (area A), reflects salt water intrusion. The presence of calcium and magnesium with considerable concentration is due to ion exchange of calcium and magnesium by sodium and long term contact and interaction between aquifer matrix and groundwater. According to the Piper diagram, groundwater in the Nile delta is classified as hard-bicarbonate water in the middle and southern parts, and alkali- non carbonate water type in the northern part.

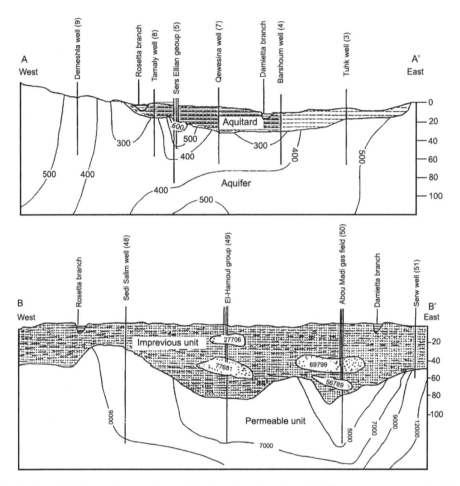

Figure 6.   Salinity (ppm) along the southern (A–A') and northern (B–B') sections crossing the Nile Delta as in Fig. 1.

12. Trace elements of strontium, bromide, boron and lithium were found in groundwater in the southern and middle parts of the study area, in concentrations closely similar to that of the Nile water (Dahab, 1994). In the northern part, these elements show high concentrations in contrast with the southern and middle parts. On the other hand, these concentrations are less than that of the sea water with the exception of some samples. This is due to reaction with the aquifer matrix.

13. Stable isotopes ratios (δD and $\delta^{18}O$) of groundwater samples collected from the southern and middle parts of the study area are closely similar to those of the Nile water. This indicates that these parts have been recently affected by the Nile fresh water (Dahab, 1994).

14. Groundwater samples collected from the zone between Tanta city and Kafr El-Sheikh show depleted stable isotopes ratios. This shows that this type of water formed in a past humid period, when the Nile fresh water flushed the aquifer.

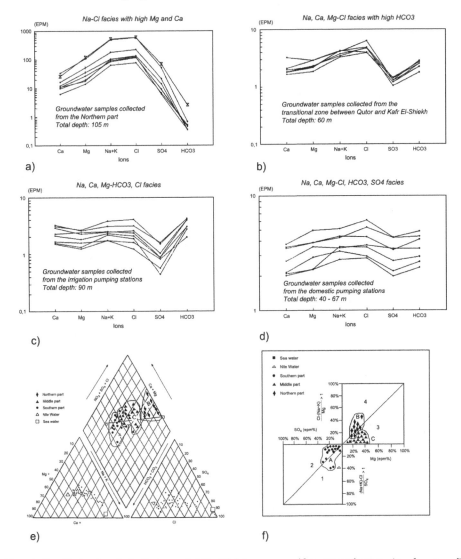

Figure 7. Hydrochemical facies of Plio-Pleistocene aquifer groundwater (a, b, c, d) and Trilinear Piper diagram (e) and Sukin diagram (f) for groundwater classification [$EPM = \frac{\text{moles of solute}}{\text{ppm}}$ valence].

15. Most of the groundwater samples collected from the northern part of the study area show enriched stable isotopic ratios compared to the present sea water. This indicates that the sea water intruded in the northern part of the study area was formed in drier periods and it is not likely connected with present sea water.

16. Depleted stable isotopic ratios in some localities in the northern part compared to the present sea water indicates heterogeneous sediments, intensive leaching and mixing by depleted south–north flux formed during the Early Pleistocene Humid periods. There are some localities where highly porous sediments are leached by a northward flux,

and others with less permeable sediments deposited under arid condition and high rate of evaporation, that prevailed during the Holocene.

Isotope and trace elements data revealed that there is also a vertical change in groundwater quality, owing to intrusion of present sea water especially in the uppermost sub-aquifer and old connate water upcoming at deep aquifer system levels (Dahab, 1994).

5   CONCLUSIONS

- Investigations carried out in the study area clearly showed that saltwater intrusion in the Nile Delta is naturally due to the hydrogeological structure of the delta sediments of different grain size deposited over geological time by the main river and its tributaries into the Mediterranean Sea, at different sea levels depending on climate changes.
- Groundwater tapped by shallow wells (15–40 m deep) in local aquifers consisting of sand and silt lenses interbedded in the Bilqas formation is mostly old connate water entrapped in sediments from an evaporitic environment, as indicated by the water table 3.5–1.5 m deeper than the present day sea level, and the hypersaline composition. These aquifers may be recharged by seepage from village waste water ponding and freshwater from the River Nile irrigation canals.
- At the uppermost levels of the Plio-Pleistocene Mit Ghamr formation aquifer system, the fresh groundwater extracted by deep drilled wells ($\geq$100 m) is recharged by the present day Nile River but at deeper levels it probably consists also of old Nile water infiltrated under humid climatic conditions.
- Fresh water is locally endangered by connate saltwater upconing and, especially in the northern part of the Nile Delta, by lateral saltwater intrusion from the Mediterranean and the coastal lakes (Idko, El-Burullus and El-Manzala).
- Salinity and major elements show higher concentrations in the northern part than in the south. Groundwater samples from the northern part, compared to sea water, are characterized by depleted salinity, chloride and sodium contents. These concentrations may possibly represent intruded sea water whose composition has been modified by cation exchange. The enrichment in calcium and magnesium indicates that the intruded sea water has been in contact with the aquifer matrix for a considerable period of time. The groundwater of the northern part is not likely connected with present day sea water.
- The natural water cycle and sedimentation processes have been strongly affected by human actions. A major impact was produced by the closure of the Aswan High Dam. With the construction of the Aswan High Dam in 1968, the river Nile sediment load was cut-off, sea erosion started attacking the delta shore, mainly in places of reduced deposition. The abatement of the great river floods strongly reduced the natural recharge of the coastal aquifer system. Therefore in the interface zone, the hydrodynamic balance between sea, surface water and groundwater with different salinity was modified, so that saltwater intrusion is probably increased, especially in the coastal area.
- At present, the balance between fresh water, sea and old saline connate water is very fragile and will rapidly worsen if groundwater exploitation increases, so that the fresh groundwater resource would be seriously endangered. Pumped groundwater yields should not exceed 2.5×10^6 m^3/year. Integrated surface water and groundwater resources of different salinity should be managed in terms of quantity and quality

for different uses to face the increasing demand in the Nile Delta, mainly for irrigation purposes.

- About 12.00×10^9 m^3/year (40% of the total surface water inflow in the Nile Delta region) are discharged into the sea and coastal lakes, through the Rosetta branch and surface and subsurface drainage network. These quantities are suitable for irrigation and should be used instead of pumping groundwater in excess.
- The present day interface zone, with a salinity of 1,000 ppm, is not deeper than 200 m in wells located in the middle part of the study area, where the suitable depth for productive wells is 100 m. In the northern part, it is around 100 m deep, so that productive wells should not be deeper than 45–50 m and be located directly beside the main irrigation canals.

REFERENCES

Dahab, K. (1994) Hydrogeological Evolution of the Nile Delta after the High Dam Construction, PhD Thesis, Geology Department Fac. of Sc., Menoufia Univ., Egypt, 164 p.

El-Menayar, M. (1999) Salt-Water Intrusion Studies on the Northern Part of the Nile Delta of Egypt, B.Sc. Thesis, Geology Department, Faculty of Science, Menoufia Univ., Egypt, 117 p.

Fairbridge, R.W. (1961) Physics and Chemistry of the Earth, 4: 99–185, Pergamon, Oxford.

Farid, M.M. (1980) The Nile Delta Groundwater Study. MSc. Thesis, Fac. of Engin., Cairo Univ., Egypt.

Otto Simonett, UNEP/GRID-Arendal, (2002). http://maps.grida.no/go/graphic/nile_delta_potential_ impact_of_sea_level_rise

Piper, A.M. (1953) A Graphic Representation in the Geochemical Interpretation of Groundwater Analysis, An. Geophys. Union Transaction, 25: 914–923.

Said, R. (1993) The River Nile: Geology, Hydrology and Utilization, Pergamon Press, Oxford, 320 p.

Saleh, M. (1981) Some Hydrogeological and Hydrochemical Studies on the Nile Delta, M.Sc. Thesis, Fac. of Sc. Ain Shams Univ., Egypt.

Schoeller, R. (1962) Les Eaux Souterraines, Masson and Cie, Paris, 642 p.

Serage El-Din, H. (1989) Geological and Hydydrogeological Studies on Quaternary aquifer PhD Thesis Fac. of Sc., Mansoura University, Egypt.

Sestini, G. (1989) Nile Delta Review of a Depositional Environments and Geological History, Egyptian Geological Society Special Publication, 41: 99–127.

Shackleton, N.J., Opdyke, N.D. (1976) Paleoceanography and Paleoclimatology Investigation of Late Quaternary, Memoir 145, R.M. Cline, J.D. Hays, Eds. (Geological Society of America, Boulder, CO, 1976), 449–464.

Siegel, F.R., Gupta, N., Shergill, B., Stanley, D., Gerber, C. (1995) Geochemistry of Holocene Sediments from the Nile Delta, Journal of Coastal Research ISSN 0749-0208 CODEN JCRSEK, 11, 2: 415–431 (1 p.1/4).

Stanley, J.D. (2005) Submergence and Burial of Ancient Coastal Sites on the Subsiding Nile Delta Margin, Egypt, Méditerranée, 1.2: 65–73.

Stanley, D.J., McRea, J.E. JR., Waldron, J.C. (1996) Nile Delta Drill Core Sample Databases for 1985–1994: Mediterranean Basin (MEDIBA) Program, Smithsonian Contributions to the Marine Sciences, 37: 428 p.

Stanley, D.J., Warne, A.G. (1998) The Nile Delta in its Destruction Phase, Journal of Coastal Research, 14: 794–825.

Warne, A.G., Stanley, D.J. (1993) Archaeology to Refine Holocene Rates along the Nile Delta Margin, Egypt, Geology, 21: 715–718.

CHAPTER 3

Impact of climate variability on the water resources in the Draa basin (Morocco): Analysis of the rainfall regime and groundwater recharge

Samira Ouysse
Dynamic of Basins and Geomatic Laboratory

Nour-Eddine Laftouhi
GEOHYD Laboratory, Geology Department, Cadi Ayyad University, Marrakech, Morocco

Kamal Tajeddine
Dynamic of Basins and Geomatic Laboratory

ABSTRACT: The High and Middle Draa basins are located in the South-East of Morocco, an area which is characterized by an arid climate. The Draa basin belongs to the 10 most arid catchments of the world. The context of the climate of the study area depends primarily on the rainfall as a principal source of the inputs to the rivers and aquifer. The mode of rainfall was studied from four upstream stations. In this study, we used several statistical methods which showed a relative homogeneity for the majority of the monitoring stations data. The mode of groundwater recharge in the Draa basin differs on both sides from the Mansour Eddahbi dam. In the High Draa, the recharge stems directly by precipitation and snow-melt. For the Middle Draa, the recharge is conditioned by the management of the reservoir dam which fluctuates according to the hydraulicity of the years. The precariousness of the water resources in the Draa basin represents clearly the main limitation for the development of this region. Under the given conditions, future changes of the climate or of the water uses will have a strong impact on groundwater availability and quality.

Keywords: Variations of climate, rainfall, statistical approaches, groundwater recharge, Draa basin (Morocco)

1   INTRODUCTION

Located in the South-East of Morocco, the High and Middel Draa constitute a quasi closed basin. The only discharge system is that of Tarhia of Draa which receives periodic releases of water from the Mansour Eddahbi dam, the majority of the inflows to which originate in the High Atlas. The aridity of the climate is due to low rainfall in opposition to high potential evaporation (Margat, 1961).

Several previous studies in the Draa basin did not discuss the quality of the existing data (Chamayou, 1966; Margat, 1961; SOGREAH, 1995). Our interest is actually founded on the quality and density of the data. Obtaining suitable input data sounds straightforward but,

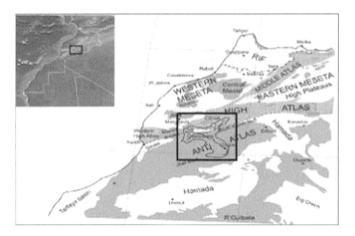

Figure 1.   Map showing the situation of the study area (Cappy 2006).

in practice, it can require care and skill. The first problem which we encountered at the time of our study was the scarcity of data, in addition to the presence of gaps in the hydro-climatic parameters which makes the analysis harder and increases the questions of data quality. This affects the majority of the monitoring stations. This study consists in subjecting the data from the principal stations (stations which had long and intact data series compared with the other stations in our study area) to some statistical tests. The statistical analysis consists on understanding the time series behaviour; testing their homogeneity and their stationarity.

2   CLIMATIC VARIABILITY AND RAINFALL DATA CONTROL

The climate context of the study area is based primarily on the rainfall, which is the primary source of groundwater and river refill. The precipitation presents the principal key element of the climate; therefore, information about trends and spatial variability of rainfall time series has become indispensable for both scientific and the practical points of view. First of all, it is necessary to make a difference between two terms which are usually confused; the climate variability and the climate change. The first one can be regarded as the variability inherent in the stationary stochastic process approximating the climate on a scale of a few decades while the last one can be regarded as the difference between the stationary processes representing climate in successive periods of a few decades (World Meteorological Organisation, 1988). The effects of climate variability can be a significant component in capturing the variability and timing of both inflow and outflow components (Hanson et al., 2004).The study of the rainfall mode was originally derived from four stations (Fig. 2) (3 upstream stations; Ait Mouted (1580 meters above sea level (m a.s.l.)), Iffre (1500 m a.s.l.) and Agouim (1647 m a.s.l.)) and one at the Mansour Eddahbi dam level (1104 m a.s.l.).

2.1   *Methods*

The data used in this study are monthly precipitation records from four principal mon-itoring stations of the Regional Direction of Hydraulics (DRH), whose time series are

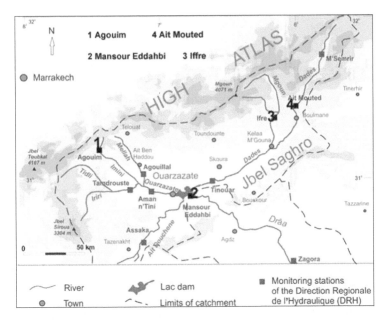

Figure 2. Locations of the four monitoring stations in the Draa basin (Cappy 2006).

Table 1. Statistical characteristics for annual precipitation at the four monitoring stations during 1975–2005.

Parameters	Stations			
	Agouim	Aït Mouted	Iffre	Mansour Eddahbi
Mean (mm)	249.4	163.7	167.6	106.4
Standard deviation (mm)	116.8	77.3	90.7	65.7

considered long enough to identify the climate signal. The technique used in this study in order to analyse the evolution of the internal structure of the time series is based on parametric and non parametric tests for testing the homogeneity and stationarity. The tool used to achieve this task is the KHRONOSTAT free software package developed by IRD-ORSTOM (*http://www.hydrosciences.org*). After testing the stationarity of the time series, we subjected these data to prediction models. The aim is to adjust the model on the data of the first years and then to predict the annual precipitation over the next years. The model used in this study is the stochastic model ARIMA (Autoregressive Integrated Moving Average).

2.1.1 *Evolution of rainfall*
The statistical study of the annual and monthly precipitation series makes it possible to know the evolution of the rainfall events. The statistical characteristics of the annual observations are given in Table 1.

The reduced centred index is the ratio of the variation in the inter-annual average to the standard deviation of the annual rainfall. It allows observation of inter-annual variability as well as the periods of pluviometric deficits and surpluses. The calculation of the annual

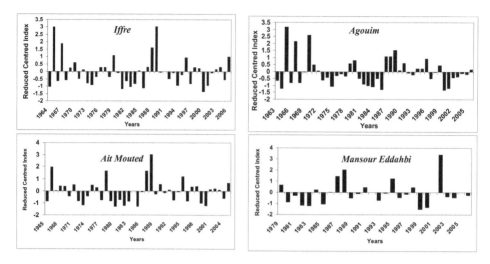

Figure 3.   Inter-annual variation in the reduced centred index rainfall for the monitoring stations (Data from DRH).

index of rainfall (Fig. 3) shows that the most marked deficit phase of precipitation was from 1981 until 1986, and this is reflected in the four monitoring stations (Ait Mouted, Iffre, Agouim and Mansour-Eddahbi). The precipitation deficit remained the general rule up to 1998, in spite of a variation in the extension and the intensity. The analysis of the inter-annual variability of the pluviometric amplitudes at the four monitoring stations shows that the greatest rise is recorded in 2002 with an average increase of +3.4 at the Mansour Eddahbi station; at the Agouim station with an increase of +3.1 in 1965; an average increase of +3 in 1988 at the Aït Mouted station; and a rise in 1989 with an average of +3 at the Iffre station.

2.1.2   Normality test
It is often useful to symmetrize the data before testing. This is important for parametric tests that are often based on the assumption of normality, but is less important for non parametric tests. Hydrological data is often highly skewed and non normal. In such cases data analysis can sometimes be assisted if the data is first transformed. The transformation of time series makes it possible to increase the normality of the data. The equation of Box-Cox transformation is presented as follow:

$$Z_p(k) = \begin{cases} \lambda_k^{-1}\left(Z_p^{\lambda_k} - 1\right) \rightarrow Si\lambda_k\rangle 0 \\ \log Z_p \rightarrow Si\lambda_k = 0 \end{cases} \tag{1}$$

where Z_p is the response variable and λ_k is the transformation parameter. The value of λ_k corresponding to the maximum correlation on the plots is then the optimal choice for λ_k.

The test of normality for the annual precipitation series shows in Fig. 4 that:

• For the Agouim station, the data follows a normal law;
• For the Aït Mouted station, the data follows a normal law after square root transformation ($\lambda_k = 0.5$);

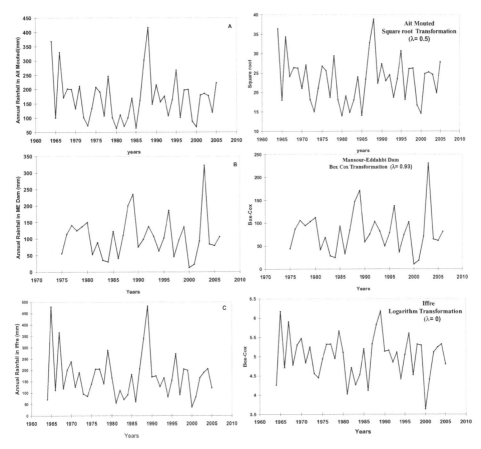

Figure 4.   Plots of Box-Cox transformation A) Aït Mouted station B) Mansour-Eddahbi station C) Iffre station.

- For the Mansour Eddahbi station, the data follows a normal law after Box Cox transformation ($\lambda_k = 0.93$);
- For the Iffre station, the data follows a normal law after logarithm transformation ($\lambda_k = 0$).

The curves before the transformations show a data set that does not follow a normal distribution; the curves of the data after applying the Box Cox transformations show a data set for which the normality assumption is reasonable.

2.1.3 *Buishand method and control ellipse*

The Buishand method (Buishand, 1982) tests the homogeneity of a time series. It is based on cumulative differences, and is effective in the case of sharp changes of the mean. By supposing a priori uniform distribution for the position of the rupture point t, the statistic U of Buishand is defined by:

$$U = \frac{\sum_{k=1}^{N-1} (S_k^*/D_x)^2}{N\,(N+1)} \qquad (2)$$

where

$$S_k^* = \sum_{i=1}^{k} (x_i - M) \quad \text{For } K = 1, \dots, N$$

D_x indicates the standard deviation of the series, and M the average. The null hypothesis of the statistical test is "the absence of rupture in the series". In the case of rejection of the null hypothesis, no estimation of the date of rupture is proposed by this test. In addition to these various procedures, the construction of a control ellipse makes it possible to analyze the homogeneity of the series of (x_i). The S_k^* variable follows a normal distribution of null average and variance $K(N - K)N^{-1}\sigma^2$, $K = 0, \dots, N$ under the null hypothesis of homogeneity of the series of x_i. Therefore, it is possible to define an area of confidence known as the control ellipse (Bois, 1971; 1986) associated to a confidence level containing the series of S_k^* under the null hypothesis. The control ellipse test is a graphical complement to the Buishand's U-Statistic. It is possible to define a confidence region containing, for a given confidence level in the null hypothesis, the S_k^* series. For a given confidence level $(1 - \alpha/2)$, the confidence region is defined by:

$$\pm \frac{U_{1-\frac{\alpha}{2}}\sqrt{k\,(N-k)}}{\sqrt{(N-1)}}D_x \tag{3}$$

where U in the normal standardized variable. This confidence region is called control ellipse.

For the four stations (Fig. 5), the hypothesis of absence of rupture was accepted at the 99%, 95% and 90% confidence levels. The Bois's Ellipse shows that $Q = \max (S_k^*)$ is located at the confidence level of 99%, 95% and 90% which implies that the hypothesis of stationarity is accepted (stationarity of the time series). To ensure these results; the stationarity of the series must be checked by the test of Pettitt.

2.1.4   *Pettitt's test*

The test of Pettitt is a non parametric approach derived from the Mann and Whitney test allowing the identification a breaking point in a sequence of random and independent variables. The Pettitt change point test (Pettitt, 1979) is used to test the occurrence of abrupt change. The test is particularly useful when no hypothesis can be made about the location of the change point. It is given as:

$$K = \max_{1 \le K \le N} |U_k| \tag{4}$$

where U_k is equivalent to a Mann–Whitney statistic for testing whether those two samples (x_1, x_2, x_θ) and $(x_{\theta+1}, x_{\theta+2}, \dots, x_{\theta+N})$ come from the same population (Demaree & Nicolis, 1990). U_k is calculated from:

$$U_k = 2 \sum_{i=1}^{k} M_i - k\,(N+1) \tag{5}$$

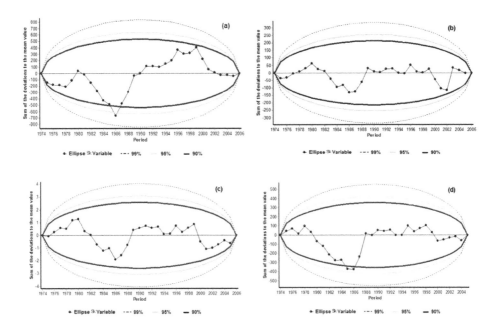

Figure 5. Buishand and Control ellipse test for the data stations (a) Agouim (b) Mansour-Eddahbi (c) Iffre and (d) Aït Mouted.

where M_i is the rank of the *i*th observation when the values $x_1, x_2, \ldots, x_N$ in the series are arranged in ascending order. A change point occurs in the series where U_k attains a maximum. To test for the statistical significance of the change point, the calculated value of K is compared with its theoretical value at probability level α, given as:

$$K_\alpha = \left[-\ln \alpha \left(N^3 + N^2 \right) / 6 \right]^{1/2} \qquad (6)$$

where a significant change point exists, the series are segmented at the location of the change point. The homogeneity and stationarity of the rainfall records of the Iffre, Aït Mouted, Agouim and Mansour-Eddahbi stations were accepted at all confidence levels 99%, 95% and 90% (Fig. 6).

2.1.5 *Bayesian statistics*

Bayesian methods have a number of attractive features. They provide a measure of the uncertainty of an estimate of change. However, they can be complex to apply and require distributional assumptions to be made (Lee, 1997). As opposed to the point estimators (means, variances) used by **classical statistics**, **Bayesian statistics** is concerned with gen-erating the posterior distribution of the unknown parameters given both the data and some prior density for these parameters. As such, Bayesian statistics provides a much more com-plete picture of the uncertainty in the estimation of the unknown parameters, especially

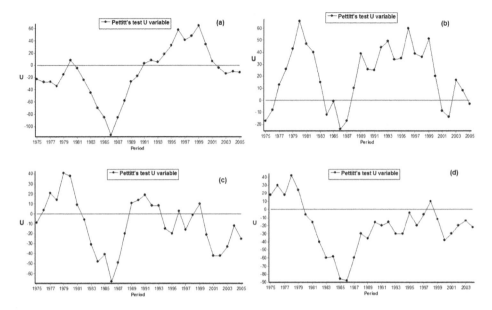

Figure 6.   The annual change in Pettitt's test U variable for the data stations (a) Agouim (b) Mansour-Eddahbi (c) Iffre and (d) Aït Mouted.

after the confounding effects of nuisance parameters are removed (Walsh, 2002). The basic model of the procedure is as follows (Kotz et al.,1981):

$$X_i = \begin{cases} \mu + \varepsilon_i & i = 1, \ldots, \tau \\ \mu + \delta + \varepsilon_i & i = \tau + 1, \ldots, N \end{cases} \tag{7}$$

The ε_i are independents and have a normal distribution, with a zero mean and a variance σ^2. τ, μ, δ and σ are unknown parameters such as, $1 \leq \tau \leq N - 1, -\infty \langle \mu \langle \infty, -\infty \langle \delta \langle \infty, \sigma \rangle 0$ $\tau, \sigma, \mu, \delta$ are independent. τ and δ represent respectively the position in time and the scope of a possible change in the mean. The Bayesian approach presented here is based on the *a posteriori* marginal distributions of τ and δ (Lee & Heghinian, 1977)

The *a priori* distributions of τ and δ are:

$$p(\tau) = 1/(N - 1) \quad \tau = 1, 2, \ldots, N - 1 \tag{8}$$

$p(\delta)$ is a normal variable with a zero mean and a variance of $\sigma \delta^2$

$$p(\delta) = \frac{1}{\sigma_\delta \sqrt{2\pi}} \exp^{\frac{-\delta^2}{2\sigma_\delta^2}} \quad -\infty \langle \delta \langle +\infty \tag{9}$$

The *a posteriori* distribution of τ is defined by:

$$p\langle \tau | X \rangle \propto [N/(\tau(N - \tau))]^{1/2} [R(\tau)]^{-(N-2)/2} \quad 0 \leq \tau \leq N - 1 \tag{10}$$

where

$$R(\tau) = H(\tau) \Big/ \sum_{i=1}^{N} (X_i - \bar{X}_N)^2$$

$$= \left[\sum_{i=1}^{\tau} (X_i - \bar{X}_\tau)^2 + \sum_{i=\tau+1}^{N} (X_i - \bar{X}_{N-\tau})^2 \right] \Big/ \sum_{i=1}^{N} (X_i - \bar{X}_N)^2$$

with

$$\bar{X}_N = 1/N \sum_{i=1}^{N} X_i \quad \bar{X}_\tau = 1/\tau \sum_{i=1}^{\tau} X_i \quad \bar{X}_{N-\tau} = 1/(N-\tau) \sum_{i=\tau+1}^{N} X_i$$

Knowing that;

$$p\langle \delta \mid X \rangle = \sum_{\tau=1}^{N-1} p\left(\delta \mid \tau, X\right) p\left\langle \tau \mid X \right\rangle \tag{11}$$

The *a posteriori* conditional distribution of δ in relation to τ, $p(\delta|\tau; x)$ is a Student's distribution; With a mean of $\hat{\delta}_\tau = \bar{X}_{N-\tau} - \bar{X}_\tau$ and a variance $\sigma^2_{\delta|\tau} = NH(\tau)/[(N-2)(\tau(N-\tau))]$ of with $\nu = N - 2$ degrees of freedom.

The Student's density function is as follows:

$$p\langle \delta \mid \tau, X \rangle = \frac{\nu^{\nu/2} \Gamma\left((\nu+1)/2\right)}{\Gamma(1/2)\,\Gamma(\nu/2)\langle \sigma_\delta \mid \tau^2 \rangle^{1/2}} \frac{1}{\left(\nu + (\delta - \hat{\delta}_\tau)^2/\sigma_\delta \mid \tau^2\right)^{(\nu+1)/2}} \tag{12}$$

The time of the break and its range are equal to the values of the modes of the *a posteriori* distributions of τ and δ respectively; probabilities are associated with the mode.

The Bayesian approach of Lee and Heghnian (1977) tested in our study indicates that the probable dates of a change on the stations time series are presented as follows in Fig. 7:

- A posteriori probability density function mode of break point position 0,1440 in 1986 for Agouim station;
- A posteriori probability density function mode of break point position 0,0954 in 1985 for Aït Mouted station;
- A posteriori probability density function mode of break point position 0,1015 in 2005 for Iffre station;
- A posteriori probability density function mode of break point position 0,0791 in 1975 for Mansour-Eddahbi station.

The Bayesian method uses the mode of the posterior distribution to estimate the parameters of the model. These results represent the posterior distribution of the break date.

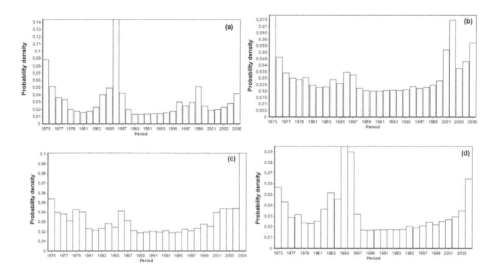

Figure 7.   Lee and Heghnian procedure for the four monitoring stations—A posteriori probability density of a break time position (a) Agouim (b) Mansour-Eddahbi (c) Iffre and (d) Ait Mouted.

The mode of the distribution is the highest probability date which occurs in the year; 2005 with 10% posterior probability of a break in the Iffre data series, 1975 with 7.9% posterior probability of a break in the Mansour Eddahbi data series, 1985 with 9.5% posterior probability of a break in the Aït Mouted data series, and 1986 with 14% posterior probability of a break in the Agouim data series.

2.1.6   *Test of segmentation*

Some classical tests, (Pettitt, 1979; Buishand, 1982) help in detecting a possible change point of the mean so that the original nonstationary series can be split into two stationary sub series. The Bayesian procedure defined by Lee and Heghinian (1977) supposes the *a priori* existence of a change of the mean somewhere in the series and yields at each time step an a *posteriori* probability of mean change. The principle of the Hubert's segmentation procedure (Hubert et al., 1989) is to cut the series into (m) segments so that the mean calculated for the whole segment is significantly different from the mean of the neighbouring segment (s). Such a method is appropriate for the research of multiple changes of the mean in a hydro-meteorological series. The segmentation procedure was first applied (Hubert et al., 1989) to West-African hydrometeorological series (33 rainfall series plus Senegal and Niger discharge). It appears to be a useful robust tool for preliminary analysis needed at the present stage of studies into climate variability and change (Hubert, 2000). The segmentation is defined as follows:

Any series X_i, $i = i_1$, i_2 with $i_1 \geq 1$ and $i_2 \leq N$ where ($i_1 < i_2$) is a segment of the initial series (X_i), $i = 1, \ldots N$.

Any partition of the initial series into (m) segments is a type m segmentation of that series.

$$i_k, K = 1, 2, \ldots, m$$

$$n_k = i_k - i_{k-1}$$

$$\bar{X}_k = \frac{\sum_{i=i_{k-1}+1}^{i=i_k} X_i}{n_k} \tag{13}$$

$$D_m = \sum_{k=1}^{k=m} d_k \quad \text{with } d_k = \sum_{i=i_{k-1}+1}^{i=i_k} \left(X_i - \bar{X}_k\right)^2$$

The segmentation retained at the end of the implementation of the procedure must be such that for a given m type segmentation; the quadratic deviation D_m is minimal. This condition is necessary but not sufficient to determine the optimal segmentation. We should add the constraint which states that the means of the two contiguous segments must be significantly different. This constraint is met by applying the Scheffé test (Dagnélie, 1970). According to Hubert et al. (1989), this segmentation procedure can be considered as a test on the stationary nature of the series, "the studied series is stationary" being the null hypothesis for the test. If the procedure does not produce an acceptable segmentation superior or equal to 2, the null hypothesis is accepted.

With Scheffe's test level of significance at 1% (strong evidence against H_0), the condition according to which the mean of the contiguous segments are significantly different; the application of the segmentation method over thirty year's (1975 to 2006) shows that the time series are stationary in all stations (Table 2). When testing this method for a period less than 30 years, the datasets show only one segment for Aït Mouted, Iffre and Mansour Eddahbi stations. In contrast, the Agouim station time series are subdivided in two segments: the first from 1971 to 1986 presenting an average of 209,66 and the second from 1987 to 1998 presenting an average of 330,82 (Table 3). These results impose the problem of time series record length (quantitative aspect). Data series should be as long as possible. Short data series can be strongly affected by climate variability which can give misleading results. For

Table 2. Hubert's segmentation for the four monitoring stations during 1975–2006.

	Beginning	End	Mean	Standard deviation
Agouim	1975	2006	250.79	115.15
Ait Mouted	1975	2006	163.69	77.34
Iffre	1975	2006	171	91.29
M.E. Dam	1975	2006	106.44	64.67

Table 3. Hubert's segmentation test for Agouim station during 1971–1998.

Beginning	End	Mean	Standard
1971	1986	209.66	97.32
1987	1998	330.82	98.36

investigation of climate change, a minimum of 50 years of record is suggested—even this may not be sufficient (Robson, 2000).

2.1.7   *The rank correlation test*

This method used a nonparametric test to check the randomness of a series. It can be used to check the stationary character of the series but it is not always true. The rank correlation test (Kendall & Stuart, 1943) is used to test the independence of the consecutive elements of a series. The null hypothesis H_0 is "the series of X_i, $i = 1, \ldots, N$ is random" where X_i indicates the realizations of the variable X observed at equal steps of time. The alternative hypothesis H_1 of this test is that of the tendency in the observations series. This test is based on the calculation of the number P of pairs (X_i, X_j) for which $(X_j > X_i)$ $(j > i, i = 1, \ldots, N - 1)$. Under the null hypothesis (H_0) of stationarity of series, the variable τ is defined by:

$$\tau = 1 - \frac{4Q}{N\,(N-1)} \quad \text{with } Q = \frac{N\,(N-1)}{2} - P \tag{14}$$

τ follows a normal distribution of null average and variance equal to:

$$\sigma_\tau^2 = \frac{2\,(2N+5)}{9N\,(N-1)} \tag{15}$$

If the null hypothesis H_0 is true, the variable $U = \tau/\sigma_\tau$ is a normal reduced variable. For a α risk of first given group, the acceptance area of the null hypothesis is included between $-U_{1-\alpha}/2\sigma_\tau$ and $U_{1-\alpha}/2\sigma_\tau$. The alternative hypothesis is the existence of a trend. When one takes immediate interest in the asymptotic distribution of the P variable, this test is called the Mann–Kendall test.

This test shows that the whole of the time series in the four stations are random, with a difference in the computation variable value, as a result:

- Agouim: computation variable value $= 0.1964$
- Iffre: computation variable value $= 0.3568$
- Aït Mouted: computation variable value $= 0.6629$
- ME Dam: computation variable value $= -0.6487$

2.1.8   *Test of correlation*

In order to understand the correlation between the variables, we have performed two types of correlation. The first correlation is a bi-dimensional correlation between two variables; its quality is measured by the dispersion of the point around the average. This test shows a prevalence of the convergent pairs with a positive development and strong correlation between (Iffre & Agouim), (ME Dam & Ait Mouted) which are strongly arranged along the straight regression line. In fact the correlation is less good between (ME Dam & Iffre), (ME Dam & Agouim); and there is no correlation between (Iffre & Ait Mouted) and (Ait Mouted & Agouim). Regarding the geographical view point, we notice that both Ait Mouted and Iffre stations, which are almost at the same elevation and the same position (East part of the basin crossed by the most important streams (M'goun & Dades) of the Draa River), are not correlated. The best correlation appears for the series of variables of (Iffre & Agouim). These results are not coherent and the stations showed unpredictable and uncertain data series.

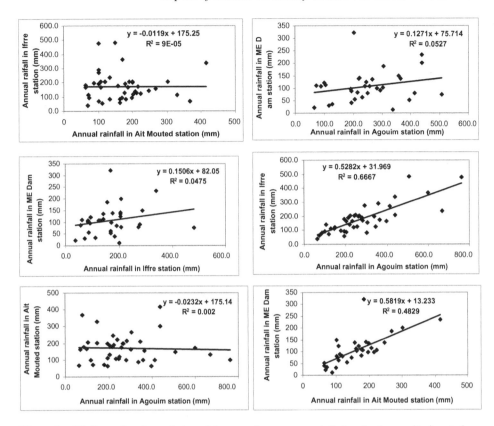

Figure 8. Bi-dimensional correlation of the annual average precipitation for the monitoring stations (Data from DRH).

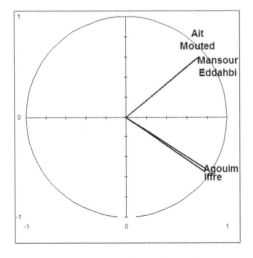

Figure 9. Principal component analysis for the four stations.

The second correlation test is the principal component analysis (correlation between variables); this test shows in Fig. 9 a big correlation between (Agouim & Iffre) and (Aït Mouted & Mansour Eddahbi) data stations determined by the first axis.

2.2   *Discussions*

For the studied period, the application of the statistical tests of detection of the randomness and existence of rupture showed that the whole of the time series at the four stations are random, homogeneous and stationary. The tests for detection of breaks proved their power and they are based on the research for one or several (Hubert segmentation) roughest changes of average in the studied series. The break is generally defined by a change of the probability law of the chronological series at the given moment, which is mostly unknown. For Pettitt, Buishand and Bois ellipse tests; the results consist of the determination of the levels of rejection or acceptance of the null hypothesis. The segmentation test of Hubert tries to detect various segments having various averages inside the same series by application of the Scheffé test. The Bayesian method gave us dates when the probability of break is the most important and it offers an exact statistical tool for inference in small samples. The posterior distributions encapsulate all the uncertainty about the parameters. Therefore the inference is exact. Other methods that use large sample theory approximate the distribution of the parameters when the sample size is large. In small samples the approximation can be very poor. Bayesian methods offer a good alternative if the sample size is small. The climatic variability at the observed stations showed a positive and relative correlation on a two-dimensional scale; the principal components analysis showed two sets of data series that area well correlated (Aït Mouted & Mansour-Eddahbi) and (Iffre & Agouim) which are defined by the first axis. According to the perceptibility of statistical tests to the length of the data series, the results of these statistical methods present an approximation of the climatic variability process in this studied region.

3   FORECASTING DATA

Time series prediction is the process of forecasting a future measurement by analysing the patterns, trends, the relationships of past and current measurements. Precipitation is the primary input quantity to the hydrological system, for this reason it will be necessary to estimate or forecast the magnitude of this important hydrological variable. The forecasting model used in this study is the Box- Jenkins stochastic models such as the Autoregressive Integrated Moving average (ARIMA) widely used in the framework of time series analysis. The models obtained by the Box-Jenkins method are not structural models which do not take into account the explanatory variables influence on explained variables. The purpose of the application of this conceptual study is to identify in which measure the precipitation which is usually considered as purely unpredictable (random) variable does not contain parts which can be predictable. The methodology requires that the studied chronological series should be stationary. This condition means that the average of the series and its variability must be finite and constant. In other term, the stationarity hypothesis is equivalent to suppose that the generative mechanism of the process is invariant in time. If the original series are not stationary on average, it is necessary to proceed by calculations of simple or seasonal differences to infer its stationarity, if the series are not stationary in variance, it will be necessary to transform it into logarithms.

This model ensured the combination of three temporal processes: Autoregressive process, integrated process and Moving average process. In general, the time series could be modelled as the association of a tendency, a seasonal variation, ARIMA and a residual. In statistics, an ARIMA model is a generalisation of an autoregressive moving average or ARMA model. These models are fitted to time series data either to better understand the data or to predict future points in the series. ARIMA models are, in theory, the most general class of models for forecasting a time series which can be stationarized by transformations such as differencing and logging. In fact, the easiest way to think of ARIMA models is as fine-tuned versions of random-walk and random-trend models: the fine-tuning consists of adding lags of the differenced series and/or lags of the forecast errors to the prediction equation, as needed to remove any last traces of autocorrelation from the forecast errors. Random-walk and random-trend models, autoregressive models, and exponential smoothing models (exponential weighted moving averages) are all special cases of ARIMA models. A non-seasonal ARIMA model is classified as an ARIMA (p, d, q) model, where:

- **p** is the number of autoregressive terms,
- **d** is the number of non-seasonal differences,
- **q** is the number of lagged forecast errors in the prediction equation.

To identify the appropriate ARIMA model for a time series, you begin by identifying the order(s) of differencing needing to stationarize the series and remove the gross features of seasonality, perhaps in conjunction with a variance-stabilizing transformation such as logging or deflating. If you stop at this point and predict that the differenced series is constant, you have merely fitted a random walk or random trend model (Recall that the random walk model predicts the first difference of the series to be constant, the seasonal random walk model predicts the seasonal difference to be constant, and the seasonal random trend model predicts the first difference of the seasonal difference to be constant, usually zero). However, the best random walk or random trend model may still have auto-correlated errors, suggesting that additional factors of some kind are needed in the prediction equation.

Before fitting the ARIMA model on the Aït Mouted data series, we stabilized the series variability using Box Cox transformation ($\lambda = 0.5$). The ARIMA model applied in our series is the ARIMA (0, 1, 1) model which seems to be appropriate to remove the trend effect and the yearly seasonality of the data; therefore, it represents a useful tool to examine the climate variability. The prediction model equation is:

$$\widehat{Y}(t) = \mu + Y(t-1) - \theta\, e(t-1) \tag{16}$$

where $e(t-1)$ denotes the error at period $t-1$; θ is the coefficient of the lagged forecast error, it is conventionally written with a *negative* sign for reasons of mathematical symmetry. The trajectory of the long-term forecasts is typically a sloping line (whose slope is equal to μ) rather than a horizontal line. The adjustment results of the ARIMA (0, 1, 1) (0, 1, 1)^{12} model which seems suitable to consider, at the same time, the trend component and the annual cyclicity observed are presented in the Fig. 10.

The model was validated over a period of 12 years to evaluate the compatibility of the model results with those observed. The same time length was used to predict the next period. The result shows that there is no forecast for the thirteen first values, because of constraints related to the model. Following this graphic the ARIMA model tested is relatively adjusted with the observed data. We perceive a smooth variation of ARIMA

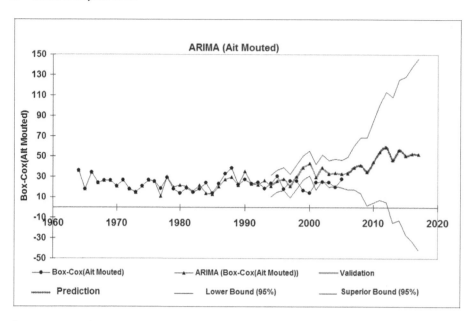

Figure 10.   Results of the ARIMA model applied to the Aït Mouted data series.

curve around the observed variable. But for the validation period the differences were much greater and for the future forecasts; the model suggest an ascending curve of the future events. The validated and predicted values are located between the two extremes of the confidence level.

4   HYDRODYNAMIC BEHAVIOUR OF THE AQUIFERS

4.1   *Groundwater behaviour system*

The study of the behaviour of the aquifers located downstream from the Mansour Eddahbi dam has been carried out on two sites (461/64 and 240/64) located at the Mezguita oasis (Fig. 11). The piezometric level data were recorded by the water resources service of Ouarzazate. This important hydrogeological parameter which informs us about the aquifer's hydro-dynamism (surface-water and groundwater exchange) reveals a database showing many gaps on a monthly scale. Gaps were infilled by the average of the closest values. Because in this study we concentrated on the rainfall parameter, the variation of the piezo-metric level in both piezometers was correlated with the rainfall variation of the nearest station (Agdz station) at the monthly scale.

The fluctuation of the groundwater level at the two sites does not show a linear rela-tionship with rainfall variations, as precipitation does not constitute the only element of aquifer recharge. Indeed the releases of water from the reservoir have an important role in the increase in the subsurface and aquifers reserves and for artificial recharge. In fact, 50% of the alluvial aquifers recharge is ensured by the irrigation surpluses.

The mode of the aquifers recharge differs on both sides from the reservoir lake, thus on the two wells located in the upstream part of the basin (Fig. 12), the fluctuation of

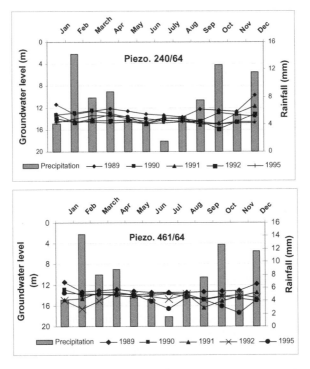

Figure 11. Variation of the groundwater level recorded at two sites in the Mezguita aquifer compared with the monthly average precipitation at the Agdz station (Data from DRH).

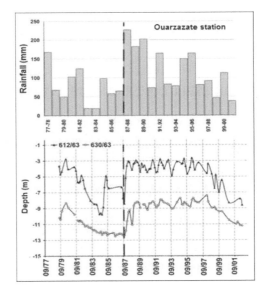

Figure 12. Groundwater fluctuations of wells in the Ouarzazate basin compared to the precipitation (inter-annual mean) for 1977 to 2001 (Cappy, 2006).

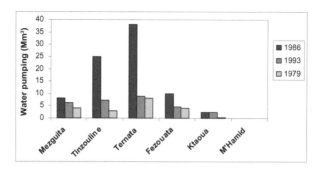

Figure 13.   Pressure of pumping during the last years presenting the different hydro-climatic conditions (Data from DRH).

groundwater is influenced by the mean variations of the inter-annual precipitations which are recorded at the Ouarzazate station.

4.2   *Groundwater overexploitation*

Water demand in the semi arid areas increased enormously during the 30 last years as a result of a succession of dry years and especially during 1980 which was marked by an exhaustive aridity in Morocco. In the Draa basin, water constitutes an important element in the subsistence of the oases. Admittedly modernization contributed to the sustainable development of the Middle Draa basin, but this was to the detriment of the Drâa River which constitutes a survival element of this zone. Today, the irrigation of the oases and the recharge of the aquifers are conditioned by the dam release which depends on the hydraulicity (wetness) of the year.

With increasing water demand, the use of another source of freshwater resource (groundwater) became a necessity. The pressure on groundwater increased strongly and shows remarkable peaks during the years of drought. In Fig. 13, 1979 presents a wet year, 1986 a dry year and 1993 an average year.

5   EVALUATION OF VEGETABLE COVER BY SATELLITE IMAGE PROCESSING

Remote sensing has become operational in the development of mapping and of environmental studies. Remote sensing is particularly adapted to the monitoring of vegetation in the visible field—Near infrared. To make a particular analysis, it is necessary to follow the evolution in the time of the whole of the components of the earth surface. In arid and semi-arid regions, vegetation covers only a small proportion of the soil surface for most of the time.

The satellite images used in this study are obtained by the ASTER sensor system on-board the Terra a satellite launched in December 1999. ASTER monitors at a spatial resolution of 90 to 15 meters. The multispectral images include near infrared, short wave infrared and thermal infrared wavelengths (*http://www.satimagingcorp.com/satellite-sensors/aster.html*).

These images are exploited in their Tiff/Geotiff format (Projection: UTM, Zone 29&30; Spheroid: WGS 84 North; Pixel size: 15.0; Unit: meters; Geo Model: Map Info).

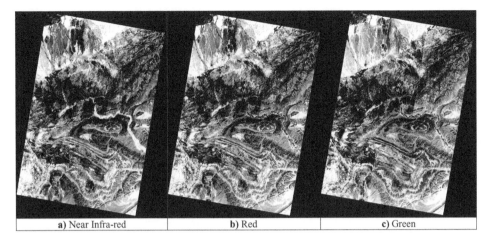

| **a)** Near Infra-red | **b)** Red | **c)** Green |

Figure 14.   Gray scale images.

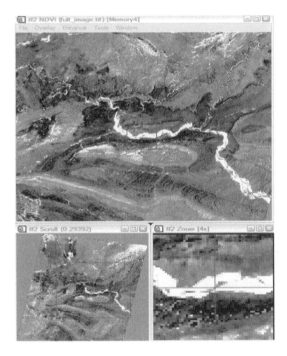

Figure 15.   NDVI image.

The visualization and analysis of the rasters were made using the ENVI (the Environment for Visualizing Images) (*http://www.ittvis.com*) image processing system. Firstly, the raster images were assembled into a single composite image using the mosaicing process. The mode of visualization in gray scale shows only a single band. The gray scale varies between 0 and 255; 0 corresponds to the black and 255 to the white. The darker a pixel, the weaker its reflectance. In Fig.14, the visualization of the image in three bands; Near infra-red (790–890 nm), Red (610–680 nm), and Green (500–590 nm) shows that the vegetation in the Draa catchment along the river valley appears in pale tints in the Near infra-red and conversely in the two others spectral bands.

The calculation of *Normalized Difference Vegetation Index (NDVI)* which is a correlated index with photosynthetic activity of land covers and which emphasizes the contrast between soil and vegetation shows an important reflectance centred in the oases fields (Fig. 15).

For the majority of continental surfaces, the NDVI varies between 0 and 1, and increases with the density of vegetation with important photosynthetic activity. The darker the pixel appears, the lower is the reflectance. This index is calculated according to the following formula:

$$NDVI = (PIR - RED) / (PIR + R) \tag{17}$$

In climatic variability studies, the vegetable cover estimation is important in the evaluation of several parameters including evapotranspiration; because this parameter is often poorly represented in our hydrological balance assessments in spite of its importance.

6   CONCLUSIONS

The Draa basin is subjected to unfavourable conditions affecting the environmental state of the area. The assessment of the climate variability is important for understanding the spatial and temporal variation in the water resources (river, groundwater). Large variability in the climate creates constraints and perturbations to our ecosystem. The most obvious features of climate variability are the strong decrease in rainfall and its spatial and temporal heterogeneity with marked periods of droughts, erratic river flows, falling water tables and high evapotranspiration rates. The analysis of the time series makes it possible to identify the parameters most affected by the climatic variations. Statistical analysis also constitutes a basic element before any step of modelling which will make it possible to realize future predictions for the various hydro-climatic factors. The results of statistical analysis of the rainfall data series did not identify a rupture, a tendency or heterogeneity of the studied data. These results cannot be radical; because during the application of the Hubert segmentation for a period less than 30 years, the results differed categorically, and that will be certainly the same case for the other tests. Therefore, it is advantageous to have a long data series record (more than 50 years), for making a concrete and more tangible vision with the climatic variation problem. Also it is necessary to take account of the other climatic parameters and mainly "evapotranspiration" which is very significant in this semi arid basin. Evapotranspiration is a very important element in the calculation of the hydrological measurement; however this parameter is often calculated by the Thornthwaite method which is based on the correlation between the monthly average temperature and the monthly Potential Evapotranspiration, which can be far from reality because we can not really estimate the percentage of the volumes of water transpired by the vegetation. The use of the remote sensing facilitates the estimate of the vegetative land cover fraction, its water requirements and the calculation of agricultural water stress. The statistical techniques and the stochastic methods constitute very important analytical tools to inform the quantitative state of the water resources; However, to have an exhaustive view regarding the precariousness of the water resources, it is necessary to introduce decision support systems which incorporate several factors and components at the same time.

ACKNOWLEDGEMENTS

We gratefully acknowledge the member of the Regional office of agricultural development of Ouarzazate, the member of water resources service in Ouarzazate and the general office of hydraulics in Rabat. We express our thanks to the National Center for Scientific and technical Research (CNRST) in Rabat for its research grant. We deeply acknowledge the International Association of Hydrogeologists to accept the results of this study presented on this paper and to be a subject of an oral communication in the 36 IAH Congress Toyama (Japan) 2008. The authors would like to thank the reviewers and Associate Editors *I. Holman, M. Taniguchi* for their constructive remarks and suggestions that significantly improved the quality of this manuscript.

REFERENCES

Application of Approximation Theory and ARIMA Models. Available online. http://www.springerlink. com/index/ul65720185797802.pdf

Auterives, C. (2002) Impact du changement climatique sur la ressource en eau en région Languedoc-Roussillon. Available online. http://www.master.sduee.upmc.fr/S_ech/P_hydro/arch/memoires2002/ Auterives2002.pdf

Bois, Ph. (1971) Une méthode de contrôle de séries chronologiques utilisées en climatologie et en hydrologie. Laboratoires de Mécanique des Fluides. Université de Grenoble. "Section hydrologie". 49 p.

Bois, Ph. (1986) Contrôle des séries chronologiques corrélées par étude du cumul des résidus. Deuxièmes journées hydrologiques de l'Orstom. Montpellier. pp. 89–100.

Buishand, T. A. (1982) Some methods for testing the homogeneity of rainfall records. Journal of Hydrology, vol. 58, pp. 11–27.

Cappy, S. (2006) Hydrogeological characterization of the Upper Draa catchment: Morocco. http://hss.ulb. uni-bonn.de/diss_online/math_nat_fak/2007/cappy_sebastien/0963.pdf

Chamayou, J. (1966) Hydrology of the valley of medium Draa. Thesis of Doct. University of Montpellier. France.

Chatfield, C. (1989) The analysis of time series. An introduction. Fourth edition. Chapman and Hall. 241 p.

Cortez, P. (2004) Evolving Time Series Forecasting ARMA Models. Journal of Heuristics, 10: 415–429.

Dagnélie, P. (1970) Théorie et Méthodes Statistiques. Vol 2. Les presses agronomiques de Gembloux. 451 p.

De Jong, C., Makroum, K., Leavesley, G. (2006) Developing an oasis-based irrigation management tool for a large semi-arid mountainous catchment in Morocco. Available online. http://www.iemss.org/iemss 2006/papers/s10/186_de jong_2.pdf

Demaree, G.R., Nicolis, C. (1990) Onset of Sahelian drought viewed as a fluctuation-induced transition, Q. J. R. Meteorol. Soc., 116: 221–238.

Desbois, D. (2006) Une introduction à la méthodologie de Box et Jenkins : l'utilisation de modèles ARIMA avec SPSS. Available online. http://www-rocq.inria.fr/axis/modulad/archives/numero-34/Desbois-34/Uneintroduction.pdf

DGR, DRH of Souss Massa and Draa (2000) Study of supply in drinking water of the country populations of the province of ZAGORA.

Direction of Mines, Geology and Energy (1977) Water Resources of Morocco (Atlasic and South Atlasic Domain). Notes and Memo. Serv. Geol. Morocco, N°231.

Dupuy, J. (1969) Hydrogeology of the valley of Draa. Rapp. Inéd. MTPC / DH / DRE.

ENVI Tutorial:Introduction to ENVI. Available online. http://www.geology.isu.edu/. . ./tutorials. . ./Tutorial 02_ENVI_Intro.pdf

Hanson, R.T., Newhouse, M.W., Dettinger, M.D. (2004) A methodology to assess relations between climatic variability and variations in hydrologic time series in the southwestern United States. J Hydrol 287(1–4):253–270.

Hubert, P., Carbonnel, J.P., Cbaouche, A. (1989) Segmentation des series hydrométéorologiques. Application a des series de précipitations et de debits de l'Afrique de l'Ouest. J. of Hydrol., 110: 349–367.

Hubert, P. (2000) Segmentation. Chapter 10. Detecting Trend And Other Changes In Hydrological Data. World Climate Programme—Water. Geneva 2000. Available online. http://www.wmo.int/pages/prog/hwrp/documents/english/WCASP-51.pdf

ITT Tutoriels ENVI. Available online. http://www.ittvis.com/ProductServices/ENVI/Tutorials.aspx

Kendall, S.M., Stuart, A. (1943) The advanced theory of statistics. Charles Griffin Londres. 2ème volume, 690 p, 3ème volume, 585 p. dans l'édition de 1977.

Kingumbi, A., Bergaoui, Z., Bourges, J., Hubert, P., Kallel, R. (2001) Etude de l'évolution des séries pluviométriques de la Tunisie Centrale. medhycos.mpl.ird.fr/doc/kin.pdf

Kotz, S., Johnson, N.L., Read, C.B. (1981) Encyclopedia of statistical sciences. New York, John Wiley. vol. 1, pp. 197–205, vol. 8, pp. 157–163, vol. 9, pp. 244–255.

Lee, A.F.S., Heghnian, S.M. (1977) A shift of the mean level in a sequence of independent normal random variables-A bayesian approach. Technometrics, vol. 19, n°4, pp. 503–506.

Lee, P.M. (1997) Bayesian statistics: An introduction, 2nd ed. Arnold, London. nitro.biosci.arizona.edu/courses/EEB596/handouts/Bayesian.pdf

Lubes-Niel, H. (1994) Caractérisation de fluctuations dans une série chronologique par applications de tests statistiques—Etude bibliographique. Rapport interne, ICCARE, n°3, ORSTOM—Hydrologie.

Margat, J. (1961) Salty waters in Morocco (Hydrogeology and Hydrochemistry). Notes and Memo. N°151.

ORMVAO and Catholic University of Louvain La Neuve (2004) Management of water and rural development integrated in the Draa valley.

Pettitt, A.N. (1979) A non-parametric approach to the change point problem', Appl. Statistics, 28, 126–135.

Robson, A.J., Reed, D.W., Jones, T.K. (1997) Trends in UK floods, In: Proceedings of 32nd MAFF conference of River and Coastal Engineers, Keel University. 12 p. Available online. http://www.wmo.int/pages/prog/hwrp/documents/english/WCASP-51.pdf

Robson, A., Zbigniew, W. Kundzewicz (2000) Detecting Trend And Other Changes In Hydrological Data. World Climate Programme—Water. Geneva 2000. Available online. http://www.wmo.int/pages/prog/hwrp/documents/english/WCASP-51.pdf

Scheffé, M. (1959) The analysis of variance. Wiley, New York, 4.77 p.

Sen, Z. (1992) Standard Cumulative Semivariograms of Stationary Stochastic Processes and Regional Correlation. Available online. http://www.springerlink.com/index/J5530274151127N4.pdf

SIC (Satellite Imaging Corporation). Available online. http://www.satimagingcorp.com

Sighomnou, D. (2004) Analyse Et Redéfinition Des Régimes Climatiques Et Hydrologiques Du Cameroun: Perspectives D'évolution Des Ressources En Eau. Available online. http://www.cig.ensmp.fr/~hydro/SOU/041013SIGHOMNOU.pdf

Sogreah, Ormvao (1995) Study of improvement of the irrigation and draining systems exploitation of the Ormvao (Stage 1). Diagnosis of the actual situation; synthesis and diagnosis.

Tarhule, A., Woo, M. (1998) Changes in rainfall characteristics in northern Nigeria. Int. J. Climatol. 18: 1261–1271.

Teixeira, A. (2005) Les séries chronologiques ou séries temporelles : présentation et principes d'analyse. www.em-consulte.com/article/157110-47k

Walsh (2002) Introduction to Bayesian Analysis. Lecture Notes for EEB 596z, c°B. nitro.biosci.arizona.edu/courses/EEB596/handouts/Bayesian.pdf

WMO (1988) Analyzing long time series of hydrological data with respect to climate variability. Project description, WCAP-3, WMOTFD-No. 224, Geneva, Switzerland. Available online. http://www.wmo.int/pages/prog/hwrp/documents/english/WCASP-51.pdf

CHAPTER 4

Effects of global warming and urbanization on surface/ subsurface temperature and cherry blooming in Japan

Makoto Taniguchi & Yohei Shiraki
Research Institute for Humanity and Nature, Kyoto, Japan

Shaogeng Huang
Department of Geological Sciences, The University of Michigan, Michigan, USA

ABSTRACT: Analysis of a long-term phenological record in Osaka show that the date of first bloom of the cherry has become earlier by 0.11 days per year during the last 60 years. This is attributed not only to surface and air temperature increases due to both global warming and heat island effects, but also to subsurface temperature warming. Analyses of long term records of the dates of the first bloom of the cherry in Japan have also been made during the last 100 years. The subsurface temperature may be the better signature or indicator for the date of the first bloom of the cherry, instead of air temperature which was used in previous studies for the estimation of the first bloom of the cherry, because the subsurface temperature is the actual accumulated and integrated thermal regime.

Keywords: Subsurface temperature, surface warming, phenology, heat island effect, cherry blooming

1   INTRODUCTION

Although global warming is considered as a serious environmental issue above the ground or near the ground surface, subsurface temperatures are also affected by surface warming (Huang et al., 2000). In addition to this global warming, the "heat island effect" due to urbanization creates subsurface thermal anomalies in many cities (Taniguchi and Uemura, 2005; Taniguchi et al., 2007; Taniguchi et al., 2009). The effect of heat islands on subsurface temperature is a global groundwater issue, because increased subsurface temperature alters soil water and groundwater systems chemically and microbiologically through geochemical and geobiological reactions that are temperature sensitive (Knorr et al., 2005). The increased subsurface temperature may alter the phenology, such as the first bloom of shrubs and tree species, because the roots of the tree are deeply affected by the subsurface environment. The importance of the effects of global warming and heat islands on the subsurface environment is now recognized (Green et al., 2007), however there has been little study on the relationships between phenology and the subsurface environment.

In this study, the effects of global warming and heat island effects on pheonology, such as cherry blooming in Japan, have been evaluated by not only surface warming but also subsurface warming.

2   METHOD

The long-term record of the date of first blooming of the cherry in Japan has been analyzed. The records have been taken at 102 meteorological stations from 1953 to 2006, including that near the center of Osaka city. The Japanese Meteorological Agency define "the first blooming day" as the day in which 5–6 flowers have started to open in the targeted cherry tree in each city. The accuracy of the measurement is one day. The long term records of monthly mean air temperature in the same meteorological stations have also been used for the analysis. The monthly mean air temperature were calculated based on the average of daily average temperature. The soil temperatures were recorded at 60 meteorological stations in Japan from 1955 to 1970 at several depths from the surface. In this study, the monthly means of soil temperature at depths of 0.1 and 0.2 meter were used for the analyses, as these are the only two depths with data available for the whole of Japan.

3   CHANGE OF CHERRY BLOOMING DATE

Phenological events such as cherry blooming deeply depend on the thermal environment. Ho et al. (2006) found that the date of first cherry blooming in Seoul became earlier by 0.14 days every year during the last 80 years. They concluded that surface warming

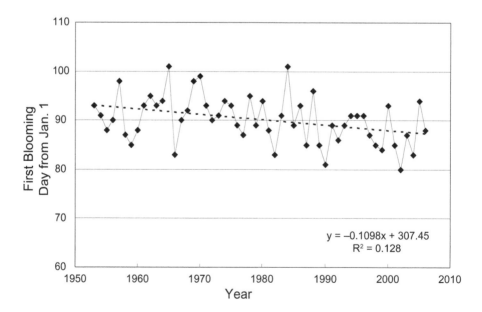

Figure 1.   First blooming day of cherry from January 1st at Osaka.

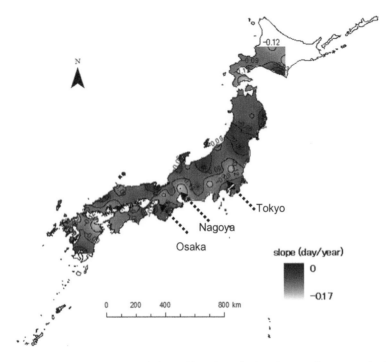

Figure 2.   Changes of the first day of cherry blooming (day/year) (Negative means that blooming becomes earlier.).

due to global warming was the cause of this phenomenon. However, the comparisons of the changes in cherry blooming have only been made with air temperature, but not with subsurface temperature. In this study, long-term comparisons between the date of cherry blooming and air temperature and soil temperature in Japan have been carried out. Fig. 1 shows the changes in the date of first cherry blooming at Osaka. As can be seen from Fig. 1, the first day of the cherry blooming becomes earlier by 0.11 days every year, although the trend is weak Osaka.

In order to evaluate the trend in the date of the cherry blooming, the same analysis has been done across Japan. The distribution of the trend of earlier blooming of the cherry is shown in Fig. 2. The trend values shows the rate at which the cherry blooming becomes earlier (i.e. −0.1 day/year means that the cherry blooming becomes earlier by 0.1 day every year). The lighter colours show the areas where the cherry blooming becomes much earlier. As can be seen from Fig. 2, the areas with lighter colours are located around Tokyo, Nagoya, and Osaka, which are three major cities in Japan. In those cities, the heat island effects due to urbanization on subsurface thermal regime have been clearly shown in the previous studies (Taniguchi and Uemura, 2005; Taniguchi et al., 2007).

4   SECULAR CHANGES IN AIR TEMPERATURE

In order to compare the cherry blooming with air temperature, more than one hundred and twenty years of records of air temperature at Osaka were analysed. Fig. 3 shows the

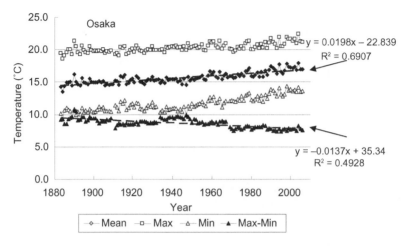

Figure 3.   Changes in annual mean, maximum, minimum, and the difference between maximum and minimum of monthly air temperature at Osaka.

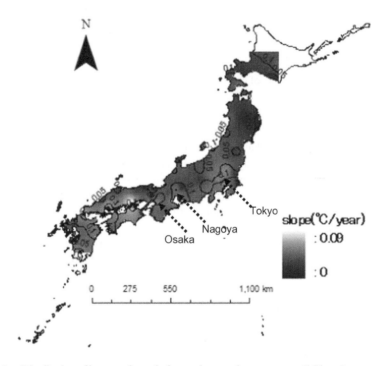

Figure 4.   Distribution of increased trend of annual mean air temperature (°C/year).

changes in annual mean, maximum, minimum, and the difference between maximum and minimum monthly air temperature.

The difference between the maximum and minimum monthly air temperature is the indicator of the heat island effect due to urbanization. The decrease in the difference between maximum and minimum monthly air temperature corresponds to the increase of heat island effect. As can be seen from Fig. 3, the annual mean air temperature increased by 0.0198°C/year at Osaka since the late 19th century. The difference between the maximum and minimum air temperature is decreasing, in particular since the 1940's.

In order to extend this analysis across Japan, the trend in increased annual mean air temperature is shown in Fig. 4. The lighter colours show the areas with the greatest trend of increasing air temperature. As can be seen from Fig. 4, the areas around Tokyo, Nagoya, Osaka and those facing the inland sea (Seto-Naikai) have the greatest trend of increasing air temperature. This distribution agrees well with the one shown in Fig. 2. Therefore, the reason for the earlier blooming of the cherry is the increase in the air temperature, in particular caused by the heat island effects due to urbanization.

5   SECULAR CHANGES IN SUBSURFACE TEMPERATURE

The relationship between the first day of cherry blooming and the air/soil temperature have been compared in Japan, using the available data across Japan. Fig. 5 shows the relationships between the first day of cherry blooming and the integrated temperature (monthly from February to April) of (a) air temperature, (b) soil temperature at 10 cm depth, and (c) soil temperature at 20 cm depth. We performed in advance the air temperature—first blooming analysis with various patterns (such as Dec–Feb, Jan–Mar, or Feb–Apr.), and found Feb–Apr. to provide the best fit for the whole of Japan. As can be seen from Fig. 5, the R^2 of those relationships is (a) 0.8545, (b) 0.9082, and (c) 0.9187, respectively. Therefore,

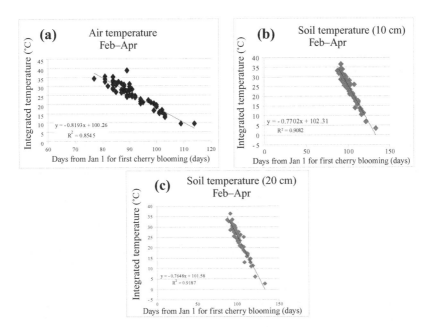

Figure 5.   Relationships between the days from January 1 for the first cherry blooming and the monthly integrated temperature from February to April of (a) air temperature (b) soil temperature at 10 cm depth, and (c) soil temperature at 20 cm depth.

the soil temperature at 20 cm depth is the best parameter to predict the first day of cherry blooming.

Only air temperature has been considered in previous studies to be a main factor controlling the phenology such as cherry blooming. The soil temperature is the integrated information of the air temperature, and the activities of roots may depend on the subsurface environment including soil temperature. This study shows that soil temperature may be a better indicator to predict the first day of cherry blooming though detail phenological studies are required in future.

6   CONCLUSIONS

The analysis of long-term records of the first day of cherry blooming in Osaka shows that the blooming has becomes earlier by 0.11 days per year during the last 60 years. Air temperature has been considered in the previous studies as a main factor controlling phenology such as cherry blooming. However the soil temperature at 20 cm depth below the surface can predict the first day of cherry blooming better than air temperature. The soil temperature is the integrated information of the air temperature, and the activities of roots may depend on the subsurface environment including soil temperature. The earlier blooming of the cherry is attributed to not only surface and air temperature increases caused by both global warming and heat island effects, but also by subsurface temperature warming. The subsurface temperature may be the better signature or indicator for the first bloom of the cherry.

ACKNOWLEDGEMENTS

The authors gratefully acknowledge the members of RIHN (Research Institute for Humanity and Nature) project "Human impacts on urban subsurface environment".

REFERENCES

Green, T., Taniguchi, M., Kooi, H. (2007) Potential impacts of climate change and human activity on subsurface water resources, Vadose Zone Journal, 6(3): 531–532.

Ho, C.H., Lee, E.J., Lee, I., Jeong, S.J. (2006) Earlier spring in Seoul, Korea, International Journal of Climatology, 26: 2117–2127.

Huang, S., Pollack, H.N., Shen, Po-Yu. (2000) Temperature trends over the past five centuries reconstructed from borehole temperatures, Nature, 403: 756–758.

Knorr, W., Prentice, I.C., House, J.I., Holland, E.A. (2005) Long-term sensitivity of soil carbon turnover to warming, Nature, 433: 298–301.

Taniguchi, M., Uemura, T. (2005) Effects of urbanization and groundwater flow on the subsurface temperature in Osaka, Japan, Physics of Earth and Planetary Interior, 152: 305–313.

Taniguchi, M., Uemura, T., Jago-on, K. (2007) Combined effects of urbanization and global warming on subsurface temperature in four Asian cities, Vadose Zone Journal, 6: 591–596.

Taniguchi, M., Shimada, J., Fukuda, Y., Yamano, M., Onodera, S., Kaneko, S., Yoshikoshi, A. (2009) Anthropogenic effects on the subsurface thermal and groundwater environments in Osaka, Japan and Bangkok, Thailand, STOTEN, doi:10.1016/j.scitotenv.2008.06.064.

CHAPTER 5

Temporal variation of stable isotopes in precipitation at Tsukuba, Ogawa and Utsunomiya City in Japan

Shiho Yabusaki
Rissho University, Magechi, Kumagaya, Saitama, Japan

Norio Tase
University of Tsukuba, Tsukuba, Ibaraki, Japan

Yasuo Shimano
Bunsei University of Art, Utsunomiya, Tochigi

ABSTRACT: Monthly precipitation samples were collected at Tsukuba, Ogawa and Utsunomiya City from 1992 to 2006, and isotope ratios of oxygen and hydrogen were determined for all samples. The isotope ratios of monthly precipitation have no remarkable trend in their seasonal change. With regard to d-excess, however, cyclic variations are observed, with relatively low d-excess values in warm periods and relatively high values in cool periods. A temperature effect on isotopes in precipitation is found during cool periods. In periods of snowfall, or precipitation from Baiu fronts and autumnal rain fronts, the isotope ratios in precipitation are relatively lighter than those of other periods. Long-term variations indicate that the annual mean air temperature is increasing and the amount-weighted mean $\delta^{18}O$ values are decreasing gradually at all sites. The annual air temperature is increasing gradually because of global warming and urbanization. The decrease in stable isotope ratio could be caused by a change of meteoric conditions for precipitation (e.g. rainfall intensity).

Keywords: Temporal variation of stable isotopes, precipitation, temperature effect, amount effect

1 INTRODUCTION

The stable isotopes of oxygen and hydrogen are useful for observing soil water movement and the groundwater flow system. Since the climate of Japan is relatively humid and recharge from precipitation is high, it is important to understand the characteristics of isotopes in precipitation for estimating water flow systems. It is also important to observe long-term variation of isotope ratios in precipitation to investigate the effects of climate change.

Dansgaard (1964) studied isotope fractionation for the stable isotopes of oxygen ($\delta^{18}O$) and hydrogen (δD) in water-vapour exchange of monthly precipitation and indicated global isotopic temperature and amount effects. The seasonal variation of stable isotopes in precipitation is affected by air temperature (temperature effect) in high latitude regions, and depends on precipitation amount (amount effect) in low latitudes. For Japan, which is located at mid latitude, the stable isotopes in precipitation are expected to be influenced by both temperature and amount effects. Yabusaki & Tase (2005) reported the temporal variation characteristics of δ^{18}O and δD in precipitation in Tsukuba City from 2000 to 2002, and confirmed that the temperature effect of isotopes in precipitation exists at Tsukuba from November to March. Machida (2000) studied spatial and temporal changes of δ^{18}O in precipitation on Miyakejima Island, and clearly showed that spatial and temporal changes in oxygen isotope ratio of precipitation is predominantly affected by amount effect and altitude effect, and also confirmed a rain shadowing effect. Yamamoto et al. (1993) observed the δ^{18}O and δD of meteoric water in Okayama Prefecture and indicated that the source of water vapour is the Pacific Ocean and the East China Sea throughout the year. Yamanaka et al. (2001) showed the time-space variation of stable isotopes in event-based precipitation on the Kanto Plain during a warm period. They showed that the most dominant pattern in both δ^{18}O and δD displays a decrease from the southeast to the northwest, particularly convective events, and that the spatial pattern may be explained by a mixing of oceanic vapour and vapour recycled in the inland area.

To estimate soil water movement and recharge rates using the vertical profile of δ^{18}O and δD in soil water, long-term isotopic data of precipitation is needed, since precipitation is considered to be the main source of soil water. In the previously described studies, however, there is little long-term isotopic data from multiple sites of precipitation, so the purpose of this study is to elucidate the characteristics of the temporal variation of stable isotopes in monthly precipitation and the relationship between air temperature or precipitation amount and isotopic ratios at Tsukuba, Ogawa and Utsunomiya City.

2   METHOD

To understand the spatial pattern and representativeness of stable isotope ratios in precipitation at the Kanto Plain, precipitation samples were collected from three sites at Tsukuba, Ogawa and Utsunomiya once a month from 1992 to 2006. A precipitation sampler, which prevents the evaporation of collected precipitation (Shimada et al. 1994), was installed on the rooftop of the Geoscience building, at the University of Tsukuba. Precipitation samplers were also installed at Ogawa City, which is located in the northwest of Saitama Prefecture, and at Bunsei Art University, Utsunomiya City, which is located in the centre of Tochigi Prefecture (Fig. 1). When the samples were collected, the amount of water was measured. In Tsukuba, the precipitation amount was calculated from the sampled water amount, and was compared with observed values at the Aerological Observatory, located in Tateno, Tsukuba, Ibaraki Prefecture, Japan (Table 1). The amount of collected water was almost equal to that of the observed values, so the amount of collected water was used as a precipitation amount. Air temperature was also measured routinely by the Aerological Observatory. Meteorological data observed at the meteorological stations in Ome and Utsunomiya were used for Ogawa and Utsunomiya, respectively (Table 1).

The isotope ratios of sampled waters were measured using the CO_2-H_2O equilibration method for oxygen, and the H_2-H_2O equilibration method with platinum as a catalyst for hydrogen. A water sample of 1 ml was put into a glass bottle and equilibrated with H_2 or CO_2 at 18°C, by shaking in a thermostatic bath for 6 hours for hydrogen, and for 9 hours for oxygen. Isotope ratios of oxygen and hydrogen were analyzed by a double collector stable isotope mass spectrometry system (Finnigan MAT 252, Thermo Electron Co., Ltd.), at the Institute of Geoscience, University of Tsukuba. Isotope ratios are expressed as

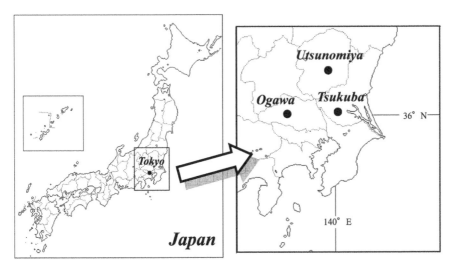

Figure 1.   Location map in this study.

Table 1.   Location of precipitation sampling points and meteorological stations.

	Sampling point	Elevation (m)	Distance from the coast (km)	Meteorological station	Elevation (m)
Tsukuba	36°06′38″N 140°06′03″E	25.0	46.2	36°03′04″N 140°07′05″E	25.2
Ogawa	36°02′49″N 139°14′25″E	98.0	122.8	35°47′03″N 139°18′07″E	155.0
Utsunomiya	36°36′23″N 139°51′18″E	158.9	70.4	36°32′09″N 139°52′01″E	119.4

the standardized permil (‰) deviation from the V-SMOW (Vienna-Standard Mean Ocean Water),

$$\delta = \left(\frac{R_{\text{sample}}}{R_{\text{SMOW}}} - 1\right) \times 1000(‰) \qquad (1)$$

where R is D/H or $^{18}O/^{16}O$. Analytical precision is ±0.1 ‰ for $\delta^{18}O$ and ±1.0 ‰ for δD. The $\delta^{18}O$ and δD values were determined for all samples of precipitation.

3   RESULTS AND DISCUSSION

3.1   *Temporal variation of monthly precipitation*

Temporal variations in precipitation amount, air temperature, $\delta^{18}O$, δD and d-excess in precipitation from 1992 to 2006 at Tsukuba, Ogawa and Utsunomiya City are shown in Fig. 2, Fig. 3 and Fig. 4, respectively. The d-excess value is defined by $d = \delta D - 8\delta^{18}O$.

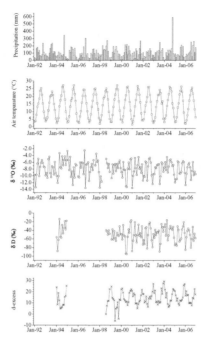

Figure 2.   Temporal variation of precipitation, air temperature, δ^{18}O, δD and d-excess in monthly precipitation at Tsukuba.

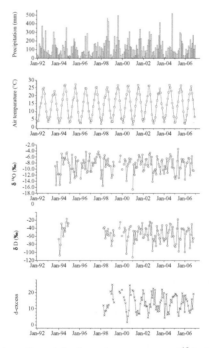

Figure 3.   Temporal variation of precipitation, air temperature, δ^{18}O, δD and d-excess in monthly precipitation at Ogawa.

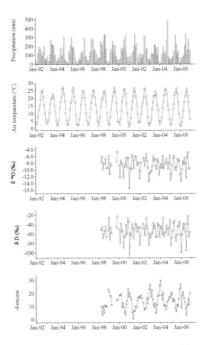

Figure 4.   Temporal variation of precipitation, air temperature, δ^{18}O, δD and d-excess in monthly precipitation at Utsunomiya.

The average annual precipitation amount is 1271 mm at Tsukuba, 1499 mm at Ogawa and 1485 mm at Utsunomiya. The mean air temperature is 14.0°C at Tsukuba, 14.1°C at Ogawa and 14.0°C at Utsunomiya (Table 2). From May to July and September to October, the precipitation amount is largely due to a Baiu front (a seasonal stationary front which developed from middle of May to beginning of July in Japan) and an autumnal rain front, respectively. The precipitation amount on October 2004 is particularly large (592 mm at Tsukuba, 521 mm at Ogawa and 490 mm at Utsunomiya) due to a typhoon. This value corresponds to the third highest rainfall on record from January 1921 to May 2008 in Tsukuba, the second highest rainfall on record from December 1976 to May 2008 in Ogawa, and the fourth highest rainfall on record from January 1891 to May 2008 in Utsunomiya.

Temporal variation in δ^{18}O is similar to that of δD. There is almost no annual trend of δ^{18}O and δD values. The amount-weighted mean values of δ^{18}O and δD in Tsukuba are −7.9‰ and −51‰, respectively. The amount-weighted mean value of δ^{18}O in Ogawa is −8.6‰ and in Utsunomiya is −8.2‰ (Table 2). In Tsukuba, the amount-weighted mean values of δ^{18}O and δD from April to September are −7.9‰ and −52‰, respectively, and from October to the following March are −8.1‰ and −47‰, respectively. During the rainy season from June to July, the isotope ratios are relatively lighter than those in other periods. In particular, isotope ratios in September of 1996 and June and July of 2000 are very low, due the development of an autumnal rain front and a Baiu front, respectively, which brought intensive rainfall. The isotope ratios are also low in January of 1998, 2001 and 2006. In January of 2001, record-breaking low air temperatures were observed in Japan, and snow fell abundantly. There was also a lot of snowfall in January of 1998 and 2006. It is thought that the stable isotope ratios in January of 1998, 2001 and 2006 were light because the

Table 2. Average monthly meteorological data (P-precipitation amount; AT-air temperature), $\delta^{18}O$, δD and d-excess at Tsukuba, Ogawa and Utsunomiya (1992 to 2006).

		J	F	M	A	M	J	J	A	S	O	N	D	Average
Tsukuba	P(mm)	54.9	39.4	98.8	95.3	131.6	123.2	149.0	114.0	170.2	167.9	81.0	45.2	1271.0
	AT(°C)	3.0	3.9	7.3	12.8	16.9	20.4	24.3	25.5	22.0	16.3	10.7	5.0	14.0
	$\delta^{18}O$(‰)	−10.2	−9.8	−8.6	−6.1	−7.0	−8.5	−8.1	−6.7	−7.1	−6.9	−6.9	−8.2	−7.9
	δD(‰)	−72	−61	−51	−35	−47	−64	−56	−48	−42	−41	−38	−45	−51
	d-excess	25.5	18.7	16.3	14.4	10.5	6.9	8.4	8.7	11.5	13.8	18.4	21.8	14.6
Ogawa	P(mm)	56.7	38.9	87.7	92.3	120.3	155.1	207.3	218.3	219.6	183.5	78.4	39.3	1499.0
	AT(°C)	3.0	4.0	7.1	12.9	17.1	20.7	24.6	25.5	21.8	16.1	10.5	5.4	14.1
	$\delta^{18}O$(‰)	−12.0	−11.4	−10.0	−7.6	−7.2	−9.0	−8.8	−6.9	−9.0	−6.3	−5.6	−8.7	−8.6
	δD(‰)	−79	−71	−62	−46	−45	−66	−59	−50	−40	−48	−38	−53	−55
	d-excess	19.8	16.9	16.1	14.3	9.6	8.0	7.9	8.7	12.4	16.0	18.8	20.1	14.1
Utsunomiya	P(mm)	42.8	34.1	87.4	108.8	161.7	161.5	224.1	188.5	214.5	154.0	73.1	35.1	1485.0
	AT(°C)	2.7	3.7	7.0	12.8	17.2	20.8	24.6	25.6	22.0	16.4	10.5	4.9	14.0
	$\delta^{18}O$(‰)	−11.1	−10.4	−8.6	−7.4	−6.8	−9.0	−8.8	−7.4	−6.7	−8.4	−7.6	−9.4	−8.2
	δD(‰)	−68	−64	−51	−46	−44	−63	−63	−50	−42	−53	−41	−53	−53
	d-excess	20.4	18.6	17.8	13.2	10.1	9.2	7.2	9.1	12.5	14.1	19.1	21.7	14.4

isotope ratios of snow samples are relatively lighter than those of rainfall samples (Moser & Stichler 1980; Lambs, 2000; Hashimoto et al., 2002).

The d-excess values have a clear seasonal variation with relatively low values from April to September (warm period) and relatively high values from October to March (cool period) every year.

The δ-diagram for precipitation in Tsukuba, Ogawa and Utsunomiya is shown in Fig. 5a, 5b and 5c, respectively. The linear equations of the local meteoric water line (LMWL) for monthly precipitation at the three sites are given in Table 3. These regression lines almost agree with the Global Meteoric Water Line ($\delta D = 8\delta^{18}O + 10$) defined by Craig (1961). For three sites, the isotope ratios of precipitation from October to March (cool period) are distributed along a line with a slope of about 8 and a y-axis intercept of 25. Ratios from April to September (warm period) are distributed along a line with a slope of about 8 and a y-axis intercept of 5 (Figs. 5a, 5b, 5c).

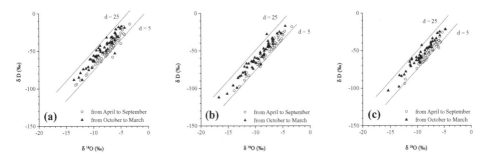

Figure 5.   Plot of $\delta^{18}O$ versus δD in precipitation at (a) Tsukuba, (b) Ogawa and (c) Utsunomiya.

Table 3.   Local meteoric water line for Tsukuba, Ogawa and Utsunomiya.

	LMWL	R^2	Error of slope	Error of Y axis
Tsukuba	$\delta D = 7.52\delta^{18}O + 10.44$	0.881	0.13	1.1
Ogawa	$\delta D = 7.63\delta^{18}O + 10.72$	0.923	0.10	0.8
Utsunomiya	$\delta D = 7.36\delta^{18}O + 8.94$	0.881	0.13	1.2

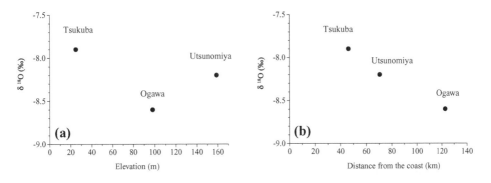

Figure 6.   Relation between (a) elevation and $\delta^{18}O$ and (b) distance from the coast and $\delta^{18}O$.

It is recognized that the isotope ratios in Ogawa are relatively light compared to those in Tsukuba and Utsunomiya (Fig. 5b). The factors affecting it could be the precipitation, air temperature, elevation or distance from the coast. The annual mean air temperature at the three sites is almost identical and the annual precipitation amount at Ogawa and Utsunomiya is almost the same (Table 2), so these effects on the isotopes are small. The relationships between elevation and amount-weighted mean $\delta^{18}O$ and distance from the coast and amount-weighted mean $\delta^{18}O$ are shown in Fig. 6a and 6b, respectively. The good negative correlation between distance from the coast and $\delta^{18}O$ suggests the existence of a continental effect on isotope ratios.

3.2   *Relationship between precipitation amount or air temperature and $\delta^{18}O$ values in precipitation*

The relationship between annual mean air temperature and amount-weighted annual mean $\delta^{18}O$ in monthly precipitation at Tsukuba from 1992 to 2006 is shown in Fig. 7. This figure

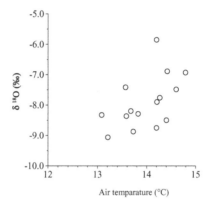

Figure 7.   Relationship between annual mean air temperature and amount-weighted annual mean $\delta^{18}O$ in precipitation at Tsukuba from 1992 to 2006.

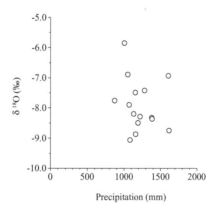

Figure 8.   Relationship between annual precipitation and amount-weighted annual mean $\delta^{18}O$ in precipitation at Tsukuba from 1992 to 2006.

indicates that there is a weak positive correlation between air temperature and δ^{18}O values. Correlations between temperature and δ^{18}O or δD similar to that of the global scale were expected. Dansgaard (1964) established a linear relationship between surface air temperature and δ^{18}O for mean annual precipitation on a global scale. Clark and Fritz (1997) indicated that the heavy isotopes (^{18}O and D) in precipitation are depleted as decreasing temperature drives rainout processes. However, our correlation of δD with temperature is poor because individual weather patterns, storm tracks and air mass mixing are too chaotic to develop a clear air temperature-δD relationship at the local scale.

The relationship between annual precipitation amount and amount-weighted annual mean δ^{18}O in monthly precipitation at Tsukuba from 1992 to 2006 is shown in Fig. 8. There is no apparent trend between the two. The isotope ratios in precipitation decrease with increasing precipitation amount according to Rayleigh processes, but, in practice the isotope ratio in precipitation is affected not only by precipitation amount, but also by air temperature.

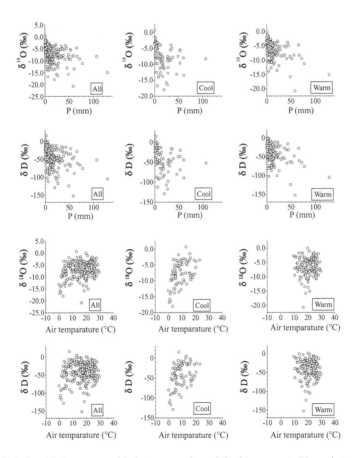

Figure 9.   Relationship between stable isotopes and precipitation amount (P) or air temperature of event data. These data are divided in all period, cool period (from October to March) and warm period (from April to September).

When the isotope ratios are divided into the warm (from April to September) and cool (from October to March) periods, as in Fig. 9, there is a weak negative correlation between precipitation amount and δD or $\delta^{18}O$ in precipitation at Tsukuba in the warm period. In general, the precipitation amount is relatively large in the warm period, and thus the influence of the amount effect on isotopic ratios is more evident in the warm period. In the case of air temperature and δD or $\delta^{18}O$ in precipitation, the positive correlation is relatively high ($r^2 = 0.36$ for $\delta^{18}O$ and 0.30 for δD) in the cool period. This is because the rainfall amount in the cool period is smaller than in the warm period. Furthermore, the correlation between air temperature and δD or $\delta^{18}O$ is shown more clearly than that between precipitation amount and δD or $\delta^{18}O$, so that it is considered that the temperature effect is more remarkable in the Kanto Plain.

3.3   *Long-term variation of annual air temperature and $\delta^{18}O$ values*

The temporal variations of annual mean air temperature and amount-weighted mean $\delta^{18}O$ in monthly precipitation from 1992 to 2006 at Tsukuba, Ogawa and Utsunomiya are shown in Fig. 10, Fig. 11 and Fig. 12, respectively. The annual mean air temperature tends to increase gradually, caused by global warming and urbanization at these sites. The rate of increase in air temperature at Tsukuba, Ogawa and Utsunomiya is 0.041°C/year, 0.048°C/year and 0.047°C/year, respectively. On the other hand, the amount-weighted mean $\delta^{18}O$ values vary with each year and tend to decrease gradually. The rate of decrease in $\delta^{18}O$ at Tsukuba, Ogawa and Utsunomiya is -0.031‰/year, -0.033‰/year and -0.052‰/year, respectively. The decrease in $\delta^{18}O$ at Utsunomiya is shown clearly. It is considered that the decrease in isotope ratio is caused by a change of meteoric conditions for precipitation (e.g. rainfall intensity). In recent years, the frequency of intense rainfall events has increased. Since the $\delta^{18}O$ or δD values of intensive rainfall events with large rainfall amounts are relatively light because of the amount effect, the isotope ratios of precipitation in recent year have become lighter in these study sites. To confirm the variation of stable isotope ratios in precipitation as global warming develops further, it will be important to continue the observation of isotope ratios in precipitation.

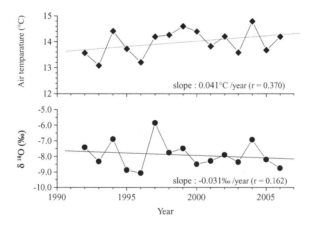

Figure 10.   Temporal variation of annual mean air temperature and amount-weighted mean $\delta^{18}O$ in precipitation at Tsukuba from 1992 to 2006.

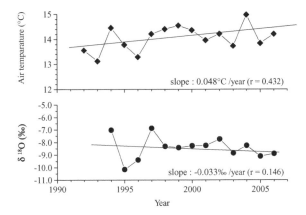

Figure 11.   Temporal variation of annual mean air temperature and amount-weighted mean $\delta^{18}O$ in precipitation at Ogawa from 1992 to 2006.

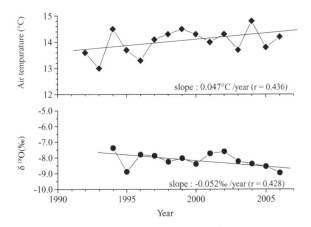

Figure 12.   Temporal variation of annual mean air temperature and amount-weighted mean $\delta^{18}O$ in precipitation at Utsunomiya from 1992 to 2006.

4   CONCLUSIONS

In this study, in order to understand the spatial pattern and representativeness of stable isotope ratios in precipitation in the Kanto Plain, stable isotopes of oxygen and hydrogen in monthly precipitation from January 1992 to December 2006 at Tsukuba, Ogawa and Utsunomiya were analyzed and the temporal variations of $\delta^{18}O$ and δD are shown. The result of this study has shown that: (1) isotope ratios of monthly precipitation have no significant seasonal cycle; (2) $\delta^{18}O$ and δD values are relatively low when Baiu fronts and autumnal rain fronts bring intensive rainfall, or during periods of snowfall; (3) the relatively low d-excess values in warm periods (from April to September) and relatively high d-excess values in cool periods (from October to March), indicated that the water vapour in warm periods was dominantly affected by the Pacific Ocean and in cool periods by the East China Sea; (4) there is a positive correlation between air temperature and $\delta^{18}O$ (temperature effect); and (5) the long-term variation in annual mean air temperature shows an increasing trend, and the long-term variation in amount-weighted mean $\delta^{18}O$ in precipitation shows a decreasing trend. It is considered that the decrease in isotope ratio is caused by a change of meteoric conditions for precipitation (e.g. rainfall intensity).

Since the intensive rainfall will be increasing gradually in the future, it seems that the stable isotope ratios in precipitation will change. Because stable isotopes are effective means of estimating water movement, it is important to grasp the isotopic variation of precipitation. And it is also necessary to continue the observation of stable isotopes in precipitation.

ACKNOWLEDGEMENTS

The authors are grateful to Professor Jun Shimada for his helpful suggestions and observations throughout this study. And we also would like to thank the reviewers for helpful advice on our paper. The research was granted a research fund from the Rissho University (Ishibashi Tanzan memorial fund).

REFERENCES

Craig, H. (1961) Isotopic variations in meteoric waters. Science Vol. 133: 1702–1703.
Clark, I., Fritz, P. (1997) Environmental isotopes in hydrogeology. Lewis Publishers, 328 p.
Dansgaard, W. (1964) Stable isotopes in precipitation. Tellus Vol. 16: 436–468.
Hashimoto, S., Shiqiao, Z., Nakawo, M., Sakai, A., Ageta, Y., Ishikawa, N., Narita, H. (2002) Isotope studies of inner snow layers in a temperate region. Hydrological Processes Vol. 16: 2209–2220.
Lambs, L. (2000) Correlation of conductivity and stable isotope ^{18}O for the assessment of water origin in river system. Chemical Geology Vol. 164: 161–170.
Machida, I. (2000) Spatial and temporal changes in oxygen isotope ratio of precipitation on Miyakejima island, Tkyo. Journal of Japan Society of Hydrology and Water Resources Vol. 13: 103–113. (*in Japanese with English abstract*).
Moser, H., Stichler, W. (1980) Environmental isotopes in ice and snow. In: Fritz, P., Fontes, J. Ch (eds.) Handbook of environmental isotopes geochemistry. Elsevier Amsterdam 1: 41–178.
Shimada, J., Matsutani, J., Dapaah-Siakwan, S., Yoshihara, M., Miyaoka, K., Higuchi, A. (1994) Recent trend of tritium concentration in precipitation at Tsukuba, Japan. Annual Report of Institute of Geoscience, University of Tsukuba Vol. 20: 11–14.
Yabusaki, S., Tase, N. (2005) Characteristics of the $\delta^{18}O$ and δD of monthly precipitation in Tsukuba from 2000 to 2002. Journal of Japan Society of Hydrology and Water Resources Vol. 18: 592–602. (*in Japanese with English abstract*).
Yamamoto, M., Kitamura, T., Akagi, S., Furukawa, T., Kusakabe, M. (1993) Hydrogen and oxygen isotope ratios of meteoric waters in Okayama Prefecture, Japan. Japanese Association of Groundwater Hydrology Vol. 35: 107–112. (*in Japanese with English abstract*).
Yamanaka, T., Shimada, J., Miyaoka, K. (2001) Time-space variation in event-based isotopic composition of precipitation over the Kanto Plain, Japan, during a warm period. Journal of Japanese Association of Hydrological Sciences Vol. 31: 123–133. (*in Japanese with English abstract*).

CHAPTER 6

The ^{14}C age of confined groundwater in a sandy-muddy Pleistocene aquifer

Isao Machida, Yohey Suzuki & Mio Takeuchi
Geological Survey of Japan, Higashi, Tsukuba, Japan

ABSTRACT: A study of the ^{14}C age of groundwater was performed along a flow path in a Pleistocene sand and mudstone aquifer system containing a small amount of carbonate. Groundwater was sampled from artesian wells varying in depth from 120 to 420 m. CaHCO$_3$ type water was detected in all samples, providing apparent ^{14}C ages ranging from 1500 to 7600 years. Geochemical and hydrological analysis indicated that the dissolved inorganic carbon was mainly from dissolution of calcite, and addition of CO$_2$ from the soil zone and organic matter in the aquifer. Based on these results, a quantitative reaction was calculated using the carbon and isotope balance method. The corrected ^{14}C ages were obtained as 100 to 3400 years with 1.3 to 3.8 m/yr of average flow rate. Despite the small calcite content in the geologic layer, the effect of dissolution on ^{14}C ages is quite significant.

Keywords: Corrected ^{14}C age, isotope mass balance

1 INTRODUCTION

Studies of the age of deep groundwater (from 100 m to 500 m depth in this paper) are rare in Japan, except for the granite Tono site. Iwatsuki et al. (2001) measured the ^{14}C of deep groundwater at this site and concluded that its age ranged from 4,000 to 19,000 years. The ^{14}C method is, in one sense, the only method available to date such old groundwater and to fill the dating range between young and very old groundwater (Kazemi et al., 2006). The application of the ^{14}C method and the accumulation of case-study data are becoming important, given the increasing need to understand the dynamics of deep groundwater.

The primary disadvantage of the ^{14}C method is the difficulty in correcting the ^{14}C age to determine the actual age, because a large number of geochemical reactions modify the concentration of ^{14}C in groundwater. Such geochemical reactions involving carbon depend on local geochemical and hydrogeological settings. In this respect, Japan has relatively recent geology, complex geomorphology and high rainfall, so that the groundwater resource is often contained in thick Quaternary (and sometimes Pliocene) systems deposited in Tertiary "vessels". Therefore, it is important to discuss the ^{14}C age correction method in Quaternary systems, particularly since Quaternary sediments often contain some marine carbonate. Although the abovementioned study at the Tono site suggests processes of groundwater chemical evolution, their study area is mainly composed of Tertiary sedimentary rocks.

This paper attempts to obtain corrected ^{14}C ages by chemical and isotope balance along a groundwater flow path in the Koito Basin, Chiba Prefecture, Japan. For a hydrogeological study, this basin has several advantages: (1) Waters can be sampled from artesian wells, which means that good quality groundwater can be sampled easily because the wells are always flushed by "fresh" groundwater. (2) It is easy to estimate the recharge zone and groundwater flow direction from the geological and geomorphological setting.

2   STUDY AREA

The Koito Basin is located midway down the Koito River, which flows through Kimitsu City, Chiba Prefecture, Japan, and has a watershed area of approximately 160 km^2 (Fig. 1). The Kimitsu City has been well characterized by the more than 200 artesian wells (Kimitsu City, 1996) contained within it. Mean annual rainfall in the region is approximately 1900 mm/yr.

The geology in and around this region is shown in Fig. 2 (Nakajima & Watanabe, 2005), and is divided into the following Quaternary geologic units in descending order: Holocene units, Shimousa Group, Kasamori Formation (Sunami sandstone Member, Sanuki mudstone Member, and Nagahama sand and gravel Member), Chonan Formation, Ichijyuku Formation, Kokumoto Formation, Umegase Formation, Otadai Formation, and Kiwada

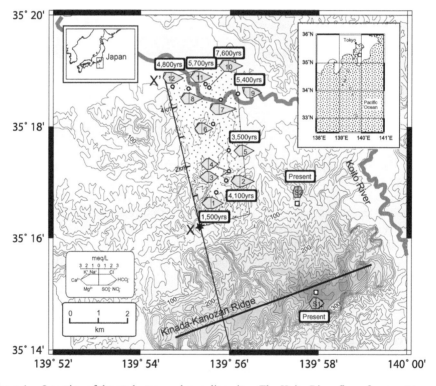

Figure 1.   Location of the study area and sampling sites. The Koito River flows from east to west. Open circles and squares indicate the sampling sites for well waters from artesian wells and springs, respectively. Site numbers appear along with Stiff diagrams, which indicate the presence of CaHCO$_3$. Apparent ^{14}C ages, which are calculated by apparent ^{14}C years = 8267 × ln(100/^{14}C$_{measured}$), are also shown.

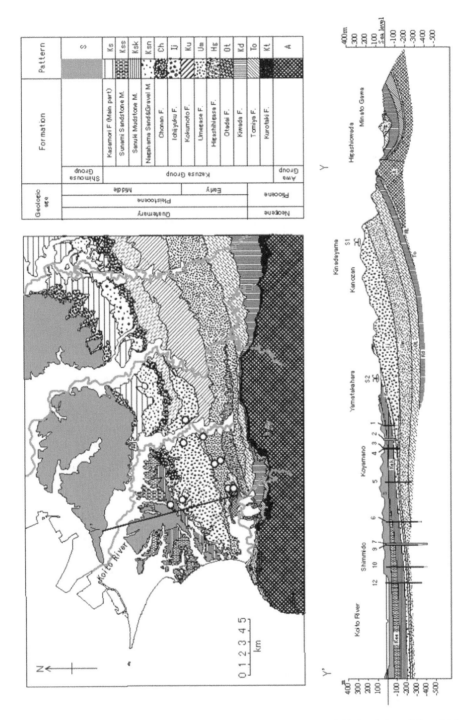

Figure 2.   Geological map (left) and geological cross section at Y-Y′ (right) (adapted from Naka-jima & Watanabe, 2005) of the Koito Basin region. The circles on the map indicate sampling points for rocks. Well numbers and depths are shown on the cross section.

Formation. While both the Nagahama Member and Ichijyuku Formation consist of unconsolidated coarse-grained sand and gravel, the Kokumoto and Umegase Formations are semi-consolidated mudstone. These geological layers are exposed at the land surface and exhibit zonal distribution.

3   HYDROGEOLOGICAL MODELS

Two hydrogeological models have been proposed for the Koito Basin. The first proposed model was presented by Chiba Prefecture (1983), and is shown in Fig. 3. The Ichijyuku Formation is distributed continuously from the Kinada-Kanozan Ridge to the Bay of Tokyo and is overlain by the Nagahama Member, which in turn is overlain by the Sanuki Member. The latter is a sandy mudstone with a low hydraulic conductivity of 10^{-8} ms^{-1}—four orders of magnitude less than that of the Nagahama Member (Chiba Prefecture, 1983). For this reason, the aquifer was assumed to consist of the Nagahama Member and the Ichijyuku Formation, and to be confined above by the Sanuki Member. The hydrological basement was taken to be the bottom of the Ichijyuku Formation. Groundwater is recharged around the Kinada-Kanozan Ridge and discharges into the Bay of Tokyo.

The second hydrogeological model was inferred from a new geological map by Nakajima & Watanabe (2005), shown in Fig. 2, and revealed that the Ichijyuku Formation decreases in thickness and is partly eroded by the Nagahama Member around the foot of the Kinada-Kanozan Ridge. According to the cross section, the aquifer system in this region should be recognized as including the sequence of the Nagahama Member to the Kiwada Formation. The hydrologic basement is unknown.

Although it is still unclear which model is correct, it is almost certain that aquiferic sediments are exposed at the southern hilly end of the Basin (Kinada-Kanozan Ridge),

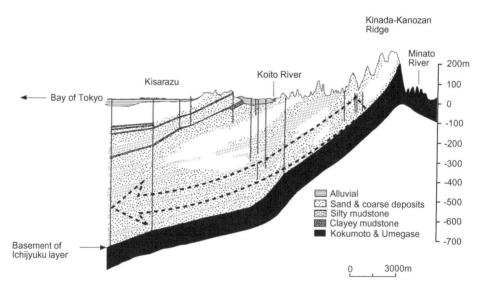

Figure 3.   Hydrogeologic model by Chiba Prefecture (modified after Chiba Prefecture, 1983), which assumes the Nagahama Member and Ichijyuku Formation as the main aquifer for deep groundwater.

where recharge occurs, and decline toward the northern lowlands. Groundwater also flows northward toward the Koito River, based on both the geological and geomorphological setting.

4 METHODS

The locations of the water and rock sampling points are shown in Fig. 1 and Fig. 2, respectively. Water temperature, pH, electrical conductivity and alkalinity were determined in the field. Other anions and cations were analyzed using ion chromatography by the Geological Survey of Japan, AIST. The $\delta^{18}O$ and δD values were determined by OPTIMA, Micromass CO. Ltd. after the preparations, CO_2–H_2O exchange technique for ^{18}O (Epstein and Mayeda, 1953) and the zinc reduction method for D (Noto and Kusakabe, 1995) in the Mitsubishi Material Natural Resources Development Corp., Saitama Prefecture, Japan. $\delta^{13}C$ and ^{14}C values of dissolved inorganic carbon (DIC) were determined by Compact ^{14}C-AMS system, NEC in Paleo Labo Co., Ltd, Saitama Prefecture. The DIC was precipitated from groundwater as $BaCO_3$ by adding $BaCl_2$ and NaOH to water samples in the field. The powdered $BaCO_3$ was converted to CO_2 gas by reacting with phosphoric acid, and the CO_2 was converted to graphite by hydrogen reduction method. The equivalent balance, Scation-Sanion, is within ± 0.2 meq/L for all samples with an analytical error of alkalinity of ± 0.05 meq/L. The analytical errors for δD, $\delta^{18}O$, $\delta^{13}C$ and ^{14}C are within ± 1‰, ± 0.1‰, ± 0.3‰, and approximately 0.3%, respectively. Rock compositions were determined by X ray diffusion analyzed by Geotechnos Co., Ltd, Tokyo Prefecture.

5 RESULTS AND DISCUSSION

5.1 *Rock mineralogy*

The results of X-ray diffusion analysis of rock samples are shown in Table 1. The Nagahama Member (Ksn), Ichijyuku Formation (Ij), Kokumoto Formation (Ku), Umegase Formation (Um) and Otadai Formation (Ot) are composed of quartz and plagioclase with small amounts of hornblende, smectite, sericite, chlorite, and calcite. Note that calcite is distributed widely in most layers. There are no significant differences between each layer. This result is quite similar to a previous report (Kashiwagi & Shikazono, 2005) which analyzed the Otadai and Ichijyuku layers.

5.2 *Human impacts on flow rate and water quality*

A comparison of the properties of the waters collected from 4 wells (No. 3, 5, 7 and 12) in 1966 and in 2008 is shown in Table 2. While the water quality shows no significant change in the past 40 years, notable decreases in flow rates are found in all four wells. The flow rates have not been monitored and the time of year of sampling in 1966 is uncertain, so that the decrease in flow rate might be derived from seasonal changes or the deterioration of the well screen. However, local residents state that the flow rates from many wells in the region have decreased over the past several decades. It is possible that this phenomenon results from reduced groundwater potential due to excess construction of new deeper artesian wells in this area.

Table 1.   Results of X-ray diffusion analysis. The double circle, circle, and triangle indicate that the mineral was abundantly, moderately, and poorly present, respectively.

	Ksn	Li	Ku	Um	Ot
Silicate minerals					
Quartz	◎	◎	◎	◎	◎
Tridymite		△		○	
Plagioclase	○	○	○	○	○
K-feldspar			△		△
Hornblende	○	△	△	△	△
Epidote	△				
Smectite	△	△		△	
Sericite/Smectite			△		△
Sericite	△	△	△	△	△
Chlorite	△	△	△	△	△
Kaolinite					
Analcime				△	
Cabonate minerals					
Calcite		△	△	△	△
Other minerals					
Pyrite				△	

The measured groundwater potential of No. 10 well (420 m depth, Fig. 1) in 2007 was +16.4 m above sea level, while the elevation of the surface of Koito River is approximately 10 m. The groundwater tends to flow from deep regions to the ground surface. It is thought that groundwater in the region has maintained its good quality despite increasing urbanization because the aquifer is deeply sourced and the groundwater has high potential.

5.3   *Validation of groundwater flow direction*

The oxygen and hydrogen stable isotopes in well waters and spring waters vary only slightly, ranging from −7.9 to −8.3‰ of $\delta^{18}O$ and −48 to −50‰ of δD. This supports that these waters have a similar recharge zone. It is also assumed that these waters were recharged after the glacial age, since there is no sign of an isotopic temperature effect between spring water (which are representative of recharging water) and artesian waters. Fig. 1 indicates that apparent ^{14}C age increases from south to north, while there are no trends from east to west (upstream to downstream in the Koito River). The distribution of these isotopes supports the notion that the groundwater generally recharges around the hilly area of the Kinada-Kanozan Ridge and flows to the north.

5.4   *Classification of groundwater quality*

The water quality of all sampled waters is represented by Stiff diagrams in Fig. 1 and all indicate the presence of $CaHCO_3$ with a range of electrical conductivity of 97 to 198 μS/cm. The Stiff diagrams have a tendency to be larger for well waters than for spring waters. On the

Table 2. Comparison of the properties of well waters collected in 1966 and in 2008. The data in 1966 were obtained by the Geological Survey of Japan (1966). EC indicates electric conductivity.

Well number	Sampling date (year)	Temp °C	pH	EC μS/cm	Flowing rate m³/day	Alkalinity (meq/L)	Cl⁻ (mg/L)	NO_2^- (mg/L)	NO_3^- (mg/L)	SO_4^{2-} (mg/L)	NH_4^+ (mg/L)	Mg^{2+} (mg/L)	Ca^{2+} (mg/L)
3	2008	14.9	8.48	113	29	1.09	5.63	0.00	0.00	9.27	0.00	2.68	17.83
	1966	15.5	8.00	105	138	1.12							
5	2008	15.5	8.51	137	58	1.35	5.36	0.00	0.00	11.65	0.00	4.65	19.71
	1966	15.7	7.80	137	92	1.44	6.40						
7	2008	16.6	8.36	198	33	1.93	6.07	0.00	0.00	10.25	0.00	5.15	21.46
	1966	16.8	8.00	143	210	1.44	8.40		0.00	12.20	0.05	5.40	19.30
12	2008	18.0	8.28	189	33	1.47	6.56	0.00	0.00	18.38	0.00	5.44	18.15
	1966	18.0	8.20	167	125	1.50	640	0.00	0.00	11.40	0.47	2.30	18.60

basis of the apparent ^{14}C ages and sampling sites, the groundwater can be classified into three groups: (A) spring waters (S1 and S2: [14]C age = present); (B) well waters at the foot of the hill (wells No. 1 to No. 5: [14]C age = 1500–4100 yrs); and (C) well waters in the vicinity of the river (wells No. 6 to No. 12: [14]C age = 4800–7600 yrs).

5.5   *Chemical evolution of groundwater in each group*

In an aquifer containing plagioclases and a small amount of calcite (Table 1), one might expect that the presence of $CaHCO_3$ in the water is mainly due to the weathering of silicates and the dissolution of calcite. Following the groundwater flow path, the springs (Group A), which are representative of recharging water, have $SI_{calcite} < 0$, while all well waters have $SI_{calcite}$ values around zero (at saturation), as shown in Fig. 4. Such variation of the saturation indices indicates that the dissolution of calcite contributes significantly to the observed water quality.

An increase in Si concentration with alkalinity is found from Group A to B, indicating the dissolution of silicates. However, a slight increase in Si concentration with increase in alkalinity can be detected from Group B to C. In this region, it is clear that the aquifer, which is located at a depth from 270 m to 420 m, is closed to soil CO_2. This suggests that the weathering of silicates is proceeded by the addition of CO_2 from the aquifer rather than the from soil zone, possibly from degradation of organic matter in the aquifer.

5.6   *Quantitative analysis for chemical evolution and correction model for* [14]*C age*

The ^{14}C ages of the well waters were corrected using isotopic and carbon mass balance concepts (Plummer et al., 1977). A trial and error method was used to adjust the differences

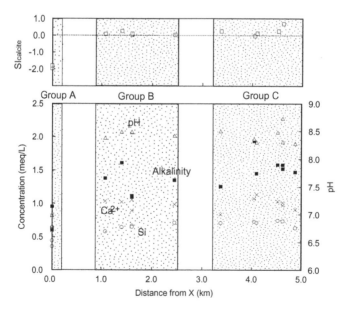

Figure 4.   Change in concentrations of ions and saturation index ($SI_{calcite}$) with respect to calcite as a function of distance from X (Fig. 1). $SI_{calcite}$ is calculated by NETPATH (Plummer et al., 1994).

in total dissolved carbon (DIC) and $\delta^{13}C$ between initial water and final water along a flow path by considering dissolution or precipitation of several phases (Table 3). Our model has three phases as possible carbon sources—calcite, organic matter in the aquifer, and soil CO_2. The differences in the corrected ^{14}C age between the initial and final water is determined by following equations:

$$\delta^{13}C_{final} = \frac{\delta^{13}C_{initial}DIC_{initial} + \Delta_{calcite}\delta^{13}C_{calcite} + \Delta_{CH_2O}\delta^{13}C_{CH_2O} + \Delta_{soilCO_2}\delta^{13}C_{soilCO_2}}{DIC_{final}}$$

(1)

$$^{14}C_{no\_decay}DIC_{final} = A_{initial}DIC_{initial} + {}^{14}C_{calcite}\Delta_{calcite} + {}^{14}C_{CH_2O}\Delta_{CH_2O}$$
$$+ {}^{14}C_{soilCO_2}\Delta_{soilCO_2}$$

(2)

$$\text{Age\_difference} = 8267 \times \ln({}^{14}C_{no\_decay}/{}^{14}C_{final})$$

(3)

where the subscripts "initial", "final", "calcite", "CH$_2$O", and "soilCO$_2$" indicate initial water, final water, calcite, organic matter in the aquifer (with dead carbon), and soil CO_2 (with active carbon), respectively. Δ is the amount of dissolution (+) or precipitation (−) in mM. DIC is calculated by NETPATH (Plummer et al., 1994). The $^{14}C_{no\_decay}$ is an estimated ^{14}C level without radioactive decay in the final water. The concentrations of $\delta^{13}C$ in calcite, CO_2 derived from organic matter in the aquifer and soil CO_2 are assumed as 0‰, −25‰, and −25‰, respectively, while corresponding ^{14}C levels are taken as 0 pmc, 0 pmc, and 100 pmc, respectively (which are the default values in NETPATH).

Equation (1) contains three unknown variables: $\Delta_{calcite}$, Δ_{CH_2O} and Δ_{soilCO_2}. The possible solutions (for $\Delta_{calcite}$, Δ_{CH_2O}, Δ_{soilCO_2}) are determined by the trial and error method taking analytical error into consideration. In this method, the calculations were done through the range from −0.2 to 2.0 mM for $\Delta_{calcite}$ and Δ_{soilCO_2} and from 0.0 to 1.0 mM for Δ_{CH_2O}. The change steps were 0.01 mM for each component. Ordinarily, the solution ($\Delta_{calcite}$, Δ_{CH_2O}, Δ_{soilCO_2}) satisfying equation (1) is not unique. In this case, unrealistic solutions can be omitted by considering hydrological and geochemical settings. For example, in the calculation of age difference between deep artesian waters from No. 5 (which is taken as the initial water for Group C, as it is the most downstream site in Group B) to No. 12 (the final

Table 3. Data for the calculations.

Sampling site	DIC	Ca^{2+}	$\delta^{13}C$	^{14}C
S2	0.77	0.30	−18.8	103.7
1	1.36	0.52	−16.7	83.4
2	1.58	0.51	−12.5	60.0
5	1.33	0.49	−13.7	64.5
9	1.44	0.57	−14.9	51.3
10	1.54	0.51	−16.1	38.6
11	1.44	0.48	−15.7	49.1
12	1.46	0.45	−14.0	55.1
	mM	mM	‰	pMC

water in Group C), the influence of soil CO_2 can be neglected because this deep region is closed to soil CO_2. Therefore, the net reaction with respect to calcite and organic matter in the aquifer must be considered to adjust the difference of DIC and δ^{13}C. In this condition, a total of 28 solutions satisfy equation (1). The dissolution amounts of calcite and organic matter range from 0.01 to 0.07 mM and from 0.05 to 0.12 mM, respectively. Waters from wells No. 5 and No. 12 are both saturated with calcite, so that the minimum dissolution amount of calcite is selected as most plausible. Consequently, it is determined as 0.01 mM. In this case, the amount of dissolution for organic matter is determined uniquely as 0.08 mM (Fig. 5). For the calculation from wells No. 5 to No. 9, we obtained three final solutions: the dissolution amount of organic matter are 0.12, 0.13, and 0.14 mM with no dissolution of calcite. In this case, the average value (0.12 mM) was selected as a plausible one. By the same procedure, the net reactions between wells No. 5 and No. 10 to No. 11 were obtained (Fig. 5).

On the other hand, the net reaction in Group (B) is profound. The calculations lead to degassing of 0.09 to 0.20 mM of CO_2 from wells No. 1 to No. 2 (in 288 solutions) and 0.15 to 0.20 mM from wells No. 1 to No. 5 (in 72 solutions). These results are unrealistic

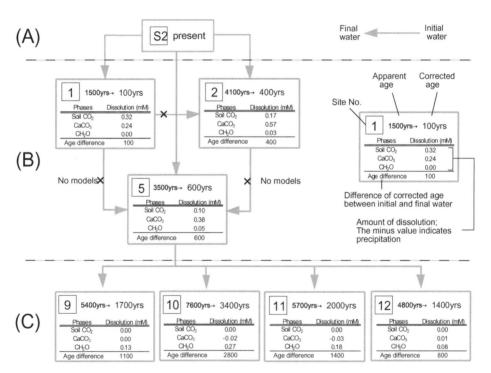

Figure 5.   Results of the calculation using equations (1) to (3). (A) to (C) indicate the water "Groups" (see section 5.4). As a calculation condition, δ^{13}C concentrations in calcite, CO_2 derived from organic matter in the aquifer, and soil CO_2 are assumed as 0‰, −25‰, and −25‰, respectively. The corresponding ^{14}C levels are taken as 0 pmc, 0 pmc, and 100 pmc, respectively. The starting point and end point of arrows indicate initial water and final water, respectively. The aim of the calculation is to fit DIC and δ^{13}C of initial and final water by trial and error method under constraints of soil CO_2, calcite, and organic matter in the aquifer. For example, DIC and δ^{13}C in well No. 5 is equal to that in a mixture of S2, 0.10 mM of soil CO_2, 0.38 mM of calcite, and 0.05 mM of organic matter.

because wells No. 1, No. 2 and No. 5 are clearly unsaturated with CO_2 and no bubbles (CO_2 gas) were observed at the well heads. Such unrealistic results may be due to very local geochemical or hydrological settings. Wells No. 3 and No. 4 are located downstream of wells No. 1 and No. 2, but they have lower electric conductivity and alkalinity (see Stiff diagrams in Fig. 1). Also, well No. 2 is located upstream of well No. 5, but the electrical conductivity and alkalinity are higher at well No. 2. Because our study basin is quite small and the groundwater has very low electrical conductivity (97 to 198 μS/cm), slight differences in geochemical and hydrological setting may significantly affect carbon chemistry.

Accordingly, it should be assumed that wells No. 1, No. 2, and No. 5 are evolved from S2 as the initial water. In such a case, the addition of soil CO_2 must be considered in equations (1) to (3) due to the existence of a vadose zone between S2 and these wells. The calculation between S2 and well No. 1 leads to 8 solutions which are similar to each other in a total of 1365 solutions in which most have unrealistic negative ages. The average of the possible 8 solutions is dissolution of 0.24 mM for calcite and 0.32 mM for CO_2 gas. Because the difference in Ca^{2+} concentration between S2 and well No. 1 is 0.22 mM, it appears that most Ca^{2+} and DIC is derived from the dissolution of calcite. However, the amount of calcite in the geology in this area is less than several % by weight (Table 1 and Kashiwagi & Shikazono, 2005), so the Ca^{2+} may be derived not only from calcite but also from silicates. It is considered, therefore, that the corrected age produced by this calculation, 100 years, is somewhat smaller than reality (Fig. 5). To calculate a more precise age, it is necessary to consider water-rock interaction, isotope exchange, measurement of carbon isotopes of soil CO_2 and calcite, and so on. As a trial, if $\delta^{13}C_{soilCO_2}$ is set as $-20‰$ in equation (1), the corrected age becomes a maximum of 700 years.

The calculations from S2 to wells No. 2 and to No. 5 lead a total of 194 and 258 possible solutions, respectively. Possible dissolution amounts of calcite range from 0.57 to 0.63 mM and from 0.38 to 0.44 mM, respectively. In this case, we select the solution with the least dissolution amount for calcite as the most plausible ones because Ca^{2+} concentrations between them change by only about 0.2 mM.

Although our calculation contains various estimations, the spatial distribution of corrected ages does not contradict the groundwater flow direction. Therefore, this paper accepts these corrected ages as a primary approximation. The corrected ^{14}C ages for waters from wells No. 1, No. 2, and No. 5 are thus calculated to be 100, 400, and 600 years, while the apparent ages are 1500, 4100, and 3500 years, respectively. For wells No. 9, No. 10, No. 11, and No. 12, their ages are obtained from the corrected age from S2 to well No. 5 plus from well No. 5 to the well in question. The differences between apparent ^{14}C ages and corrected ages range from 1400 to 4200 years. According to the horizontal distance of sampling sites from X and the corrected ages of groundwater, the average groundwater flow rate from X to wells No. 9 through No. 12 can be calculated as 1.3 to 3.8 m/yr.

6 CONCLUSION

Waters from artesian wells along a groundwater flow path were obtained from an aquifer containing a small amount of calcite. $CaHCO_3$ type water with electrical conductivity of less than 200 μS/cm was detected at all sample sites. Corrected ^{14}C ages were calculated which range from 100 to 3400 years, while apparent ages range from 1500 to 7600 years.

As a primary approximation, the corrected ages seem to be reasonable. The differences between the corrected and apparent ages are mainly due to dissolution of calcite in spite of the small calcite content in the aquifer. Our results indicate that careful attention must be paid to the existence of carbonate in aquifers in the application of the ^{14}C method.

REFERENCES

Chiba, Prefecture (1983) Chikasui tekisei riyouryou chousa houkokusyo—Kimitsu and Kisarazu—. Chiba, Japan. Chiba Prefecture (in Japanese).

Epstein, S., Mayeda, T. (1953) Variation of ^{18}O content of waters from natural source's. Geochimica et Cosmochimica Acta 4: 213.

Geological survey of Japan (1966) Chiba-ken no chikasui, geological survey of Japan Ibaraki, Japan. Geological Survey of Japan, AIST (in Japanese).

Iwatsuki, T., Xu, S., Mizutani, K., Hama, K., Saegusa, H., Nakano, K. (2001) Carbon-14 study of groundwater in the sedimentary rocks at the Tono study site, central Japan. Applied geochemistry 16: 849–859.

Kashiwagi, Y., Shikazono, N. (2005) Water-rock reaction in sequestration of carbon dioxide in sedimentary basin: a case study of Boso Peninsula, Chiba, central Japan. Journal of Groundwater hydrology 47: 65–80 (in Japanese with English abstract).

Kazemi, GA., Lehr, J.H., Perrochet, P. (2006) Groundwater age. John Wigley & Sons, Inc., Hoboken, New Jersey.

Kimitsu City (1996) Kimitsu-shi shi Shizen-hen. Daiichi-hoki, Tokyo, Japan (in Japanese).

Nakajima, T., Watanabe, M. (2005) Geology of the Futtsu Distinct. Quadrangle Series, 1: 50,000. Geological Survey of Japan, AIST, Ibaraki, Japan (in Japanese with English abstract).

Noto, M., Kusakabe, M. (1995) Reactivity of zinc metal for preparation of hydrogen isotope analysis. Technical Report of Institute for study of the earth's interior Okayama University, Ser.B, 14 (in Japanese).

Plummer, N.L. (1977) Defining Reactions and Mass Transfer in Part of the Floridian Aquifer, Water Resources Research 13: 801–812.

Plummer, N.L., Prestemon, E.C, Parkhurst, D.L. (1994) An Interactive code (NETPATH) for modeling NET geochemical reactions along a flow PATH version 2.0. Water Resource Investigation Report 94-4169, Reston, Virginia, USGS.

CHAPTER 7

Understanding groundwater flow regimes in low permeability rocks using stable isotope paleo records in porewaters

Masako Teramoto
Nippon Koei Co., Ltd., Kojimachi 4-chome, Chiyoda-ku, Tokyo, Japan

Jun Shimada
Kumamoto University, Kurokami 2-chome, Kumamoto, Japan

Takanori Kunimaru
Japan Atomic Energy Agency, Hokushin, Horonobe-cho, Teshio, Hokkaido, Japan

ABSTRACT: To investigate the regional scale groundwater flow system in low permeability siliceous Tertiary sedimentary rocks, analyses for electric conductivity (EC), hydrogen and oxygen stable isotopes (D and ^{18}O) have been done on porewaters of drilled cores. Porewaters were extracted by the compression method from eleven drillhole boreholes (HDB1-11) drilled at Horonobe, Japan. ^{18}O, D and EC data show that the origin of groundwater can be divided into three groups; present rainwater, paleo rainwater and fossil sea water. The estimated percolation depth of paleo rainwater (rainwater from during the ice age) is about 400 m below present sea level, while present rainwater has percolated only down to 160 m. The relationship between the percolation depth of rainwaters and hydraulic gradients are also investigated. The regional groundwater flow pattern estimated from the hydraulic potential distribution considerably accords with the percolation of fresh water revealed by isotopic information in porewater. These results show paleo records of regional groundwater flow would be preserved in bound water in low permeability sedimentary rock cores.

Keywords: Groundwater flow, low permeability rocks, stable isotope, paleo, porewater

1 INTRODUCTION

Groundwater flow systems in low permeability sedimentary rocks are not well understood. However they have come into the spotlight for underground facilities such as oil storage caverns and high level radioactive waste disposal vaults. This study aims to investigate the regional scale groundwater flow system in low permeability siliceous Tertiary sedimentary rocks bored for Horonobe Underground Research Laboratory (URL) Program by Japan Atomic Energy Agency (JAEA). The objective of this project is to enhance the reliability of an underground disposal of a high level nuclear waste through investigations of deep geological environment such as geology, characteristics of sedimentary rocks and groundwater flows.

It is difficult to measure hydraulic properties directly or get groundwater samples from low-permeable bed rocks. In this study, we extracted porewaters from core samples from boreholes and analysed electric conductivity, $\delta^{18}O$ and δD. From the geochemical information of porewaters, we reveal the origin of the groundwater and the flow regime.

2   STUDY SITE

Horonobe is located in the northern part of Hokkaido, and faces the Sea of Japan (Figure 1). This area is mainly formed by upper Cretaceous, Tertiary and Quaternary marine deposits. It is characterized by N-S trend faults and echelon folds. Figure 2 shows the simplified geological column of eleven borehole sites (HDB1-11). Core samples were collected

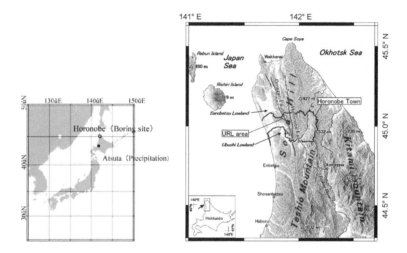

Figure 1.   Study site.

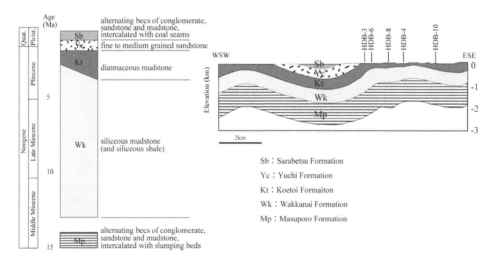

Figure 2.   Geological column and cross section.

through the Yuchi Formation (Late Pliocene to Pleistocene) which comprises coarse sediments deposited in a coastal environment; the Koetoi Formation (late Miocene to Early Pliocene) which is composed of biosiliceous mudstone; and the Wakkanai Formation (late Middle Miocene to early Late Miocene) which is composed of biosiliceous shale (Fukuzawa, 1992).

3   METHODOLOGY

3.1   *Porewater extraction*

Porewaters within drilled cores mostly exist as firmly bound water, which consists of water molecules oriented on the sediment particles so that they cannot move under natural conditions. It is possible to extract even firmly bound water by the compression method using the compaction type porewater extraction apparatus introduced by Kiho et al. (1999).

We shaved off the surface of the core samples to remove the part where pore water has evaporated and to fit the apparatus (ϕ50 mm, 100 mm length). Pore waters were extracted under the pressure of 2–70 MPa for 8–72 hours. In total, 111 porewater samples were taken (Table 1).

3.2   *Stable isotope and EC analysis*

The concentrations of stable isotopes in porewater samples were measured by an isotope ratio mass spectrometer. The concentrations of stable isotopes are expressed in the conventional delta notation in parts per thousand difference of the ratio of D to H and ^{18}O to ^{16}O in the samples relative to Standard Mean Ocean Water (SMOW) as follows;

$$\delta(‰) = \{R_{SMOW} - R_{Sample})/R_{SMOW}\} * 1000$$

where R is D/H or $^{18}O/^{16}O$. The precisions are $\pm 1.0‰$ for δD and $\pm 0.1‰$ for $\delta^{18}O$, respectively.

Electric conductivity has been measured by the compact conductivity meter (HORIBA Conductivity Meter B-173) which can measure from a single drop of porewater sample. The precisions are 2%FS $\pm$ 1digit under 1000 mS/m and 3%FS $\pm$ 1digit over 1000 mS/m.

Table 1.   The number of core samples.

Boring No.	Core sample	Boring No.	Core sample
HDB-1	12	HDB-7	5
HDB-2	11	HDB-8	4
HDB-3	9	HDB-9	10
HDB-4	9	HDB-10	12
HDB-5	10	HDB-11	23
HDB-6	6	Total	111

4   RESULTS

4.1   *δ diagram*

Isotope compositions of porewaters and shallow groundwater in the study area are displayed in Figure 3. The Local Meteoric Water Line was derived using the rain data measured at Atsuta in Hokkaido (See Figure 1). The relationship between δD and $\delta^{18}O$ of precipitation at Atsuta is $\delta D = 7.0\delta^{18}O + 6.9$. The shallow groundwater at Horonobe, which has $\delta^{18}O$ and δD of $-10‰$ and $-64.4‰$, respectively, is located on the Local Meteoric Water Line.

Isotope ratios in porewaters are distributed on a line with a slope of 3.6–4.5 from under the light proportion near the shallow groundwater to the heavy proportion. The heaviest isotope ratio values are $+4.4‰$ for $\delta^{18}O$ and $-15‰$ for δD.

4.2   *Profiles of stable isotope ratios*

Figure 4 shows the $\delta^{18}O$ and δD depth profiles of all boreholes. The isotope ratio of near surface samples in this study area have a light isotope ratio in δD that is slightly lighter than the shallow groundwater. The isotope ratios of porewaters have a trend towards heavier ratios with depth.

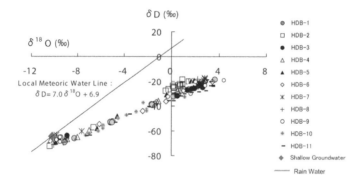

Figure 3.   Delta diagram.

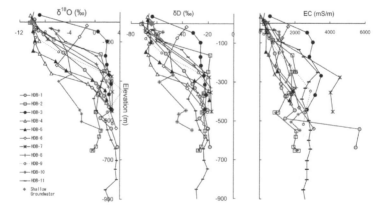

Figure 4.   Depth profiles of $\delta^{18}O$, δD and EC of porewaters.

4.3   *Electrical conductivity*

Electrical conductivity of the porewaters is shown in Figure 4. EC has a trend from low values at shallow depth to higher at deeper depths, although the values differ between the cores of the boreholes. The major dissolved ions are Na^+ and Cl^- in these high EC porewaters (Kunimaru et al., 2003).

5   DISCUSSION

5.1   *Origins of porewater*

It is revealed from the stable isotopes and EC that groundwater in this study area is a mixture of surface water and deep groundwater; surface groundwater is fresh water with light isotopes and deep groundwater is saline water with heavy isotopes and high EC. Such deep saline water is called fossil water. Its $\delta^{18}O$ is heavier than SMOW because of the isotopic fractionation with sediments (Savin and Epstein, 1970).

Figure 5 is the delta diagram of the light isotopic region in close-up. As shown in Figure 5a, linear regressions of porewater values from the cores in the HDB-1, 2, and 3 borehole cross the LMWL at the lighter ratio than the present mean shallow groundwater. This implies that freshwater with lighter isotopes than present surface water have been recharged into the sediments. This figure also shows that the origins of the freshwater differ among the three boreholes, as the crossing points of these three regressions are not the same. However, the boreholes are too close to have been recharged by different kinds of meteoric water. We recalculated the linear regressions without the porewater samples whose isotope ratios are located between LMWL and the initial regressions to remove the mixing effect of present rainwater, and got the following regressions (Fig 5b):

$$HDB\text{-}1 \; : \; \delta D = 4.118\delta^{18}O - 31.80$$

$$HDB\text{-}2 \; : \; \delta D = 4.531\delta^{18}O - 25.76$$

$$HDB\text{-}3 \; : \; \delta D = 3.951\delta^{18}O - 32.80$$

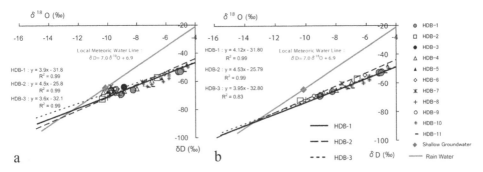

Figure 5.   Delta diagrams with linear regressions for (a) all samples and (b) excluding those whose isotope ratios are located between LMWL and the regression for all samples.

These three regressions converge at points that have similar ratios to each other as follows:

$$\text{HDB-1} : (\delta^{18}O, \delta D) = (-13.42, -87.11)$$

$$\text{HDB-2} : (\delta^{18}O, \delta D) = (-13.23, -85.68)$$

$$\text{HDB-3} : (\delta^{18}O, \delta D) = (-12.87, -83.18)$$

The average crossing point is $(\delta^{18}O, \delta D) = (-13.17, -85.32)$, which is 3‰ smaller in $\delta^{18}O$ and 10‰ smaller in δD than that of present shallow groundwater. According to the relationship between the stable isotope ratio and temperature (Hoffman et al., 2000; Rudolph et al., 1984), it is estimated that this freshwater is precipitation that occurred under 5°C cooler climate conditions than the present climate in the study area.

5.2   Ratio of present/paleo rainwater in porewater

Groundwater in this area consists of present rainwater, paleo rainwater, and fossil seawater. We calculated the proportions of present and paleo rainwater and fossil seawater using end member mixing analysis. The end members used here are $\delta^{18}O = 3.6‰$ (the heaviest sample in HDB-1), $\delta^{18}O = -13.17‰$ (the average crossing point), and $\delta^{18}O = 10.2‰$ (the shallow groundwater). Figure 6 shows the results of the calculation. It clearly shows that paleo rainwater percolates deeper than the present rainwater. The percolation depth of paleo rainwater is about 400 m below present sea level, while present rainwater has percolated only down to 160 m. It is believed that the difference in sea level between glacial and inter glacial epochs caused the diversity of groundwater flow system activities. It is also apparent that there is an inconsistency between the geological boundaries and the isotope ratios. For example, present rainwaters percolate into Wakkanai Formation in HDB-4 and 5 while there are freshwaters only in upper Koetoi Formation in HDB-1 and 3. It shows us that the regional diversity of the groundwater flow is not due to geology.

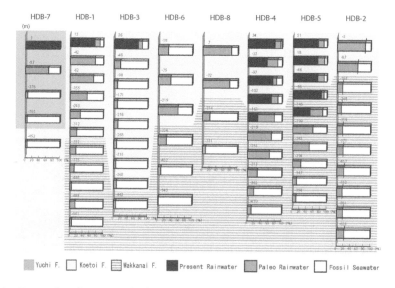

Figure 6.   Proportion of present and paleo rainwater and fossil seawater within porewater samples.

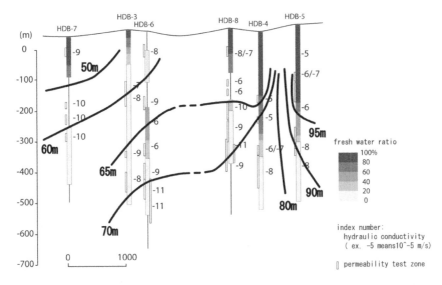

Figure 7.   Comparison of freshwater ratio and hydraulic conductivity and potential distribution.

5.3   *Comparison of freshwater ratio with hydraulic conductivity and gradients*

Hydraulic conductivities were measured by double-packer tests. Most of the sedimentary rocks in this study area are composed of extremely low permeability rocks whose hydraulic conductivity is smaller than 10^{-8} m/s. The maximum hydraulic conductivity is 10^{-5} m/s measured at shallow depths in HDB-4, 5 and 8, and the minimum is 10^{-10} m/s measured in HDB-7 and 8. Compared with the mixing ratio of freshwater (present and paleo rainwater) to saline water at certain depth and hydraulic conductivities, it is found that the percolation of freshwater is greater in relatively permeable rocks whose hydraulic conductivity is between 10^{-6} m/s to 10^{-5} m/s.

The relationship between the percolation depth of freshwater and hydraulic gradients was also investigated. Figure 7 is the E-W cross section of the study area. The shading of the cores indicate the proportion of fresh water. The index numbers of hydraulic conductivity are shown at the tested depths. Contour lines are estimated by hydraulic potential distribution. The regional groundwater flow patterns estimated from the hydraulic potential distribution agree well with the percolation evidence of fresh water revealed by the isotopic information in the porewaters.

5.4   *Comparison of isotope ratio with travel time of percolation*

To consider the validity of groundwater flow regime estimated by the isotope ratios, the percolation depth of present and paleo rainwater was compared with the results of numerical simulations. Tokunaga et al. (2006) has simulated the diffusion of fossil water and groundwater flow velocity in the Horonobe area. The results showed that the age of freshwater in sediments in HDB-5 and 7 ranges from 50 thousands to 100 thousands years, and 10 thousands years in HDB-11. The travel time of percolation revealed from the isotope ratio would be consistent with this if the paleo rainwater infiltrated during the last Ice Age (10 to 70 thousands Before Present).

6   CONCLUSION

The regional scale groundwater flow system in low permeability siliceous sedimentary rocks was investigated using the stable D and O isotopes in porewaters.

Oxygen-18 (^{18}O), Deuterium (D) and electric conductivity (EC) data show that the origin of groundwater in this study area could be divided into three groups; present rainwater, paleo rainwater and fossil sea water. The δ^{18}O of porewaters range widely from $-13.1‰$ in paleo rainwater, $-10.2‰$ in present rainwater to 3.6‰ in fossil sea water. δD ranges from $-85.3‰$ in paleo rainwater, $-64.4‰$ in present rainwater and $-18‰$ in fossil sea water. δ^{18}O and δD content in porewater samples are generally distributed from 0 to 3‰ and -20 to $-30‰$ respectively. The maximum EC is about 6000 mS/m recorded at 500 m below sea level in HDB-1. The percolation depth of paleo rainwater is about 400 m below present sea level, while present rainwater has percolated only down to 160 m. It is believed that the difference of sea level between glacial and inter glacial epochs has caused the diversity of groundwater flow systems. The boundaries of sedimentary rocks and the percolation depth of freshwater are not consistent.

Hydraulic conductivities were measured by double-packer tests. Most of the rocks in this study area are composed of extremely low permeability rocks whose hydraulic conductivities are smaller than 10^{-8} m/s. The maximum hydraulic conductivity is 10^{-5} m/s measured at shallow depths in HDB-4, 5 and 8, and the minimum is 10^{-10} m/s measured in HDB-7 and 8. The freshwater ratio and percolation depth are consistent with the hydraulic conductivity. The relationship between the percolation depth of freshwater and hydraulic gradients were also investigated. The regional groundwater flow pattern estimated from the hydraulic potential distribution considerably accords with the percolation evidence of freshwater revealed by the isotopic information in the porewater. These results of this study show that the paleo record of regional groundwater flow can be preserved in the bound water in low permeability sedimentary rock cores.

REFERENCES

Fukuzawa, H., Hoyanagi, K., Akiyama, M. (1992) Stratigraphic and paleoenvironmental study of the Neogene formations in northern Central Hokkaido, Japan. The Journal of the Geological Society of Japan. Vol. 37: 1–10.

Hoffman, G., Jouzel, J., Masson, V. (2000) Stable water isotopes in atmospheric general circulation models, Hydrological Process, Vol. 14: pp. 1385–1406.

Kiho, K., Oyama, T., Mahara, Y. (1999) Production of the compaction type pore water extraction apparatus and its application to the deep-seated sedimentary rock. Journal of the Japan Society of Engineering Geology Vol. 40. No. 5: 260–269.

Kunimaru, T., Takeuchi, R., Seo, A. (2003) Geochemical properties of groundwater in Horonobe. Paper presented at the fall meeting of the Japanese Association of Groundwater Hydrology 2003.

Rudolph, J., Rath, H.K., Sonntag, C. (1984) Noble gases and stable isotopes in 14C-dated palaeowater from central Europe and the Sahara. Isotope Hydrology 1983. IAEA Vienna: 467–477.

Savin, S.M., Epstein, S. (1970a) The Oxygen and Hydrogen isotope geochemistry of clay minerals. Geochimica et Cosmochimica Acta. Vol. 34: 25–42.

Tokunaga, T., Kimura, Y., Ijiri, Y., Honjima, T., Kunimaru, T., Takamoto, N., Shimada, J., Hosono, K. (2006) Analysis of stable chlorine isotopic ratio and saline water diffusion in Horonobe. Paper presented at the spring meeting of the Japanese Association of Groundwater Hydrology 2006.

CHAPTER 8

Mineralogical analysis of a long-term groundwater system in Tono and Horonobe area, Japan

Teruki Iwatsuki
Japan Atomic Energy Agency, Horonobe Underground Research Center, Hokushin, Horonobe-cho, Teshio-gun, Hokkaido, Japan

Takashi Mizuno & Katsuhiro Hama
Japan Atomic Energy Agency, Mizunami Underground Research Laboratory, Yamanouchi, Akeyo-cho, Mizunami, Gifu, Japan

Takanori Kunimaru
Japan Atomic Energy Agency, Horonobe Underground Research Center, Hokushin, Horonobe-cho, Teshio-gun, Hokkaido, Japan

ABSTRACT: A methodology to demonstrate the long-term hydrochemical evolution of deep groundwater is indispensable in the geological isolation of high level radioactive waste (HLW). A key component is to extrapolate the future changes in hydrochemical conditions based on the analogue of past geological events and their impact on hydrochemical conditions, and their relationships within the space of a geological system. This study conducted hydrochemical research using secondary minerals to evaluate the long-term changes in the groundwater systems in: (i) crystalline rock in the Tono area (Honshu, Japan), and (ii) sedimentary rock in the Horonobe area (Hokkaido, Japan). The spatial distribution of fracture-filling carbonate minerals provides an indication of the depths to which long-term weathering and rock-water interaction by infiltration of fresh recharge water have affected the geological sequence at each site. The petrographic characteristics (including morphology, multiple generations) and isotopic/chemical signatures of the carbonate mineralization suggest that the groundwaters have been displaced several times during their history in the Tono area. In contrast, the deep system in the Horonobe area appears to have preserved ancient groundwaters. Hydrochemical interpretations based on mineralogical information are useful to construct hydrogeological models of both areas, in particular providing insights into groundwater residence time.

Keywords: Groundwater residence time, mineralogical record, isotope, sedimentary rock, granite

1 INTRODUCTION

Understanding the long-term hydrogeological and hydrochemical evolution of groundwater systems is important to long-term water resource management and development, and in the geological isolation of high-level radioactive waste (HLW). It is particularly important for the geological isolation of HLW, in order to develop a sound scientific basis to any safety assessment for the disposal in deep geological formations. To satisfy these requirements, it is necessary to develop and test methods for understanding the long-term stability of the deep geological environment with regard to groundwater flow, and geochemical properties and processes.

Palaeohydrogeological analysis provides important insights into the long-term change of hydrogeological and hydrochemical conditions. Palaeohydrogeology is the reconstruction of past groundwater compositions, groundwater flow rates and flow directions, using a combination of geological evidence, chemical and isotopic compositions of groundwaters, mineralogical data and hydraulic measurements (Chapman and McEwen, 1993). Information on the stability of deep groundwater systems is particularly valuable for developing conceptual models of future changes in the deep sub-surface.

Evaluation of the evolution of hydrochemical conditions of a deep groundwater system is generally based on groundwater origin, residence time and water-rock interaction along groundwater flow paths. To some extent, the history of past groundwater movement can be deduced from the present chemical characteristics of sampled groundwaters (e.g. tritium, ^{14}C, ^{36}Cl activities etc.). These data may sometimes be used to deduce some general characteristics of past chemical conditions, for example, inferring past salinity from recharge age versus salinity correlations. However, groundwaters may be displaced by subsequent groundwater flushing and consequently, the groundwater chemistry may only reflect a transient state of the system. Consequently, this type of data is of limited value for evaluating past hydrochemical conditions over long time scales.

In contrast, minerals precipitated from groundwater are less easily removed by subsequent groundwater flushing events, and interpretations based on mineralogical information may potentially provide a means to consider the stability of deep groundwater over longer timescales. In particular, chemical and isotopic compositions of carbonate minerals are useful because these minerals precipitate at low temperatures in response to many processes, including the mixing of chemically distinct groundwaters, and as a result of increases of temperature along flow paths. Carbonate minerals can potentially record isotopic and chemical information that reflects the geochemical and environmental variations in the groundwaters from which they precipitated (Metcalfe et al., 1998; Milodowski et al., 1998; Iwatsuki et al., 2002; Kanai, 2007; Tullborg et al., 2008).

This study aims to develop a methodology for evaluating the long-term stability of groundwater in deep systems up to 1,000 meters below ground level (mbgl). To this end, the crystal morphology, chemistry and isotopic composition of carbonate minerals were studied, to evaluate the palaeohydrogeology of two groundwater systems: (i) in crystalline (granitic) host rock at the Tono area (Honshu, Japan) and (ii) in sedimentary host rocks at the Horonobe area (Hokkaido, Japan).

2   GEOLOGICAL SETTING AND APPROACH

2.1   *Tono area*

The Tono area is located in the central part of Honshu, Japan (Fig. 1). In this area, Plio-Pleistocene and Miocene sedimentary rocks unconformably overlie a basement of Cretaceous granite. Hydrogeological models based on topography, hydraulic head and the rock permeability suggest that groundwater flows from the north and north-east of the study area (10×10 km), towards a discharge area within the valley of the Toki River in

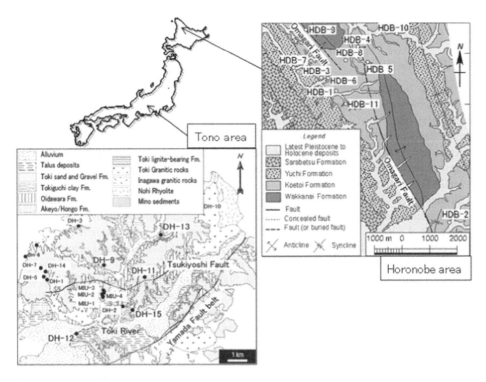

Figure 1.   Geology and borehole layout in the Tono and Horonobe areas.

the south-west (Fig. 1). The composition of the present groundwater in the granite to a depth of up to 1,000 mbgl can be divided into two types: (i) a Na-Ca-HCO$_3$ or Na-HCO$_3$ dominated type in the northern recharge area, and (ii) a fresh water of Na-HCO$_3$-Cl or Na-Cl dominated type around the discharge area (Furue et al., 2003; Metcalfe et al., 2003). The residence time of groundwater estimated by the ^{14}C content indicates that it may be several thousands of years old within the deep part of the granite (Iwatsuki et al., 2005).

Previous studies in the Tono area investigated the isotopic compositions and morphology of carbonate minerals in the basement granite up to 1,000 mbgl. These data, coupled with the known geological history of the area, suggest that three distinct fluids have flowed through the granite in the past; (i) early hydrothermal fluids; (ii) relatively saline groundwater, when the region was inundated with seawater during marine transgressions in the Miocene, and (iii) fresh water, possibly similar to modern groundwater (Iwatsuki et al., 2002; Mizuno and Iwatsuki, 2005).

Previous and new isotopic data (carbon and oxygen isotopic compositions of calcite) from boreholes (DH-9, 11, 12, 13 and 15) were used to provide a systematic profile along the line of regional groundwater flow from the recharge area to the discharge area. The spatial distribution of carbonate minerals was compared with the present hydraulic regime to evaluate the correlation between hydraulic condition and mineralogical properties. Furthermore, several carbonate mineral samples displayed different crystal morphologies within multilayered bands of mineralization. This mineral stratigraphy possibly records the chemical evolution of groundwaters along the flow path. The sequence of groundwater replacement and evolution was analyzed and inferred by using observations of the multilayered mineralization.

2.2 *Horonobe area*

The area is located in the northern part of Hokkaido (Fig. 1). This region is underlain mainly by Neogene to Quaternary marine sedimentary rocks (siliceous and diatomaceous mudstones) (Ishii et al., 2006). The area also has a widespread distribution of marine terrace deposits, which correspond to the marine oxygen isotope stages (MIS) 5c to 9 (Niizato and Yasue, 2004). The geological history of the area implies that the recharge water has possibly varied several times between fresh, brackish and seawater in the past several hundred thousand years.

Groundwater flow simulation and particle-tracking modelling suggest that the groundwater residence time in the deep groundwater system would be several million years (Niizato et al., 2008). The deep Na-Cl-HCO$_3$ dominant type groundwater, with an electrical conductivity of 40–3,500 mSm^{-1}, is inferred to be of fossil seawater origin—based on isotopic signatures (Hama et al., 2007). The groundwater residence time estimated by He and Cl radioisotopes is older than 1.5 Ma. Stable isotope analysis ($^{18}O/^{16}O$, D/H) also suggests that the groundwater system has been in a static condition for a long time at depths deeper than approximately 400 m below sea level (Teramoto et al., 2006). It may be possible to check the feasibility of using mineralogical methods to estimate the past hydrochemical conditions of the deep static groundwater system in the sedimentary rocks at Horonobe.

Carbonate mineral samples such as calcite mineralization and shell fossils were collected from rock cores taken from up to 1,000 mbgl that were drilled in the siliceous/diatomaceous mudstones. Microscopic observation and chemical/isotopic analysis of these minerals were carried out to characterize their mineralogical properties and to estimate the long-term change of groundwater from which they precipitated. Carbon and oxygen isotope ratios were determined by isotope ratio mass spectrometer. The results are reported using the conventional δ-notation, with respect to the PDB (Pee Dee Belemnite) standard for $^{13}C/^{12}C$ and the SMOW (Standard Mean Ocean Water) standard for $^{18}O/^{16}O$. The precisions are ±0.2‰ for both $\delta^{18}O$ and $\delta^{13}C$.

3   RESULTS AND DISCUSSION

3.1 *Tono area: Crystalline rock, fresh water, groundwater residence time of scale for 10 ka*

Observations of calcite fracture mineralization in the granite from depths greater than approximately 200 mbgl were made within the exposed granite area. Calcite observations from the uppermost part of the granite were made within the area where sedimentary rocks overlie the granite (Fig. 2). The boundary depths of calcite precipitation would be controlled by the long-term infiltration of recharge water and its chemical composition. The near surface recharge water is of neutral to slightly acidic pH, and is under-saturated with respect to carbonate minerals. Subsequently, it gradually evolves to become slightly alkaline and saturated to super-saturated with respect to carbonate minerals by water-rock interaction with increasing depth. The absence of carbonate minerals in the shallow part is consistent with the long-term dissolution of carbonate minerals by the recharging fresh

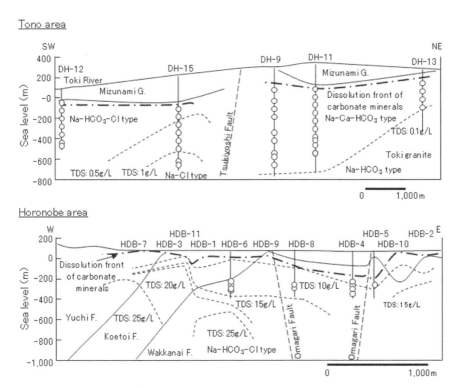

Figure 2. Spatial distribution of groundwater chemistry and carbonate minerals. Open circles are the sampling points of carbonate minerals.

groundwater. The upper limit of calcite occurrence in the granite in DH-11 at −24 m above sea level (m asl) is deeper than in DH-9 (92 m asl). The total flux of the calcite-undersaturated recharge water is greater in DH-11 than that in DH-9, and depends on the distribution of the present sedimentary cover over the granite.

The carbon and oxygen isotope ratios of calcite range from −40 to 3‰PDB and from −3 to 27‰ SMOW, respectively (Fig. 3). Previous studies classified the origins of the calcite mineralization on the basis of its isotopic ratio as follows (Iwatsuki et al., 2002):

– calcite precipitated from relatively high-temperature hydrothermal solutions ($\delta^{18}O_{SMOW}$ from −3 to c. 10‰, $\delta^{13}C_{PDB}$ from c. −18 to −7‰).
– calcite precipitated from seawater, probably partly of Miocene age ($\delta^{13}C_{PDB}$ of c. 0‰ and $\delta^{18}O_{SMOW}$ > c. 20‰).
– calcite precipitated from fresh water ($\delta^{13}C_{PDB}$ that was significantly < 0‰ and as low as c. −29‰ and $\delta^{18}O_{SMOW}$ > c. 17‰).

Calcites with $\delta^{13}C_{PDB}$ values with <−29‰ were observed down to −400 m asl of granite in DH-11 and 15 (Fig. 4). The precipitation of these calcites might be attributed to microbial process such as through bicarbonate produced by methanogenesis.

In boreholes north side of the Tsukiyoshi Fault, the occurrence of marine calcite with $\delta^{13}C_{PDB}$ of c. 0‰ (Fig. 4) tends to be restricted to DH-9, compared with DH-11 and 13

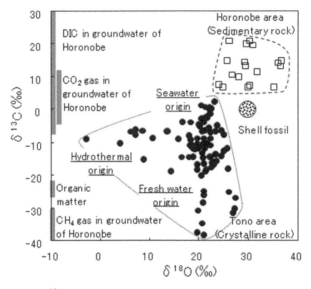

Figure 3. $\delta^{13}C_{PDB}$ and $\delta^{18}O_{SMOW}$ values of carbonate minerals at the Tono and Horonobe areas.

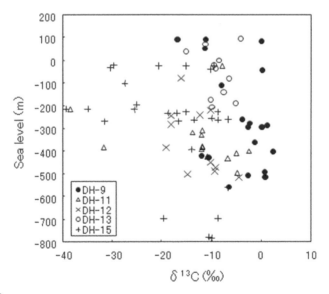

Figure 4. $\delta^{13}C_{PDB}$ values of carbonate minerals with depth at the Tono area.

located near the Tsukiyoshi Fault and in the upstream area, respectively. This area has undergone uplift within the least 5 Ma (Sasao et al., 2004). The isotopic data of calcite indicates the possibility that the marine waters were flushed by dilute freshwater recharge flowing from the northern area toward the southern discharge area and also in the vicinity of Tsukiyoshi Fault. Hydraulic head gradients driving groundwater flow to flush marine water may have persisted throughout this period. To the south of the Tsukiyoshi Fault, the difference of the $\delta^{13}C_{PDB}$ values between DH-12 and 15 are indistinct due to the paucity of marine calcite samples. Although it is not possible to analyze $\delta^{13}C_{PDB}$ because the sample amount is insufficient, the marine calcites identified by morphology are also observed at DH-15 (Mizuno and Iwatsuki, 2005).

Multilayered fracture-filling calcites, representing several generations of mineralization, were observed at DH-9 and 15. These samples were studied to assess the long-term chemical changes of flowing groundwater in the fracture flow paths. Four generations of calcite can be differentiated on the basis of petrographical observations, isotope and fluid inclusion analysis and are denoted from oldest to youngest as generations I to IV (Mizuno and Iwatsuki, 2005). The $\delta^{13}C_{PDB}$ data suggest that the Generation I calcite is of hydrothermal origin, while generations II and IV are precipitated from freshwater, and generation III is derived from seawater. Fluid inclusions were only observed in generations II and III. The fluid inclusions show no obvious change in fluid salinity between generation II and III calcite mineralization. The generation II and III calcites were possibly precipitated from mixture of freshwater and seawater. The calcite crystal morphology is sensitive to the chemistry and salinity of the groundwater. Morphological change during generations II to IV coincides with the changes of groundwater origin indicated by stable isotope and fluid inclusion composition. Based on these results, the chemistry of the groundwater had changed as a result of the flushing and the displacement of the original hydrothermal water successively by fresh water, seawater and a second episode of freshwater flushing. The depth profiles of the calcite with marine geochemical signatures suggest that replacement of groundwaters at depths below 1,000 mbgl occurred in response to marine transgression/regression.

3.2 *Horonobe area: Sedimentary rock, saline water, groundwater residence time of scale for 1 Ma*

Carbonate minerals occur in several forms in the sedimentary rocks of the Horonobe area, including; fossil shells, vein carbonate and diagenetic carbonate concretions. Owing to the paucity of open fractures in these mudrocks, multilayered calcite showing good crystal forms is very uncommon. However, several samples show the crystal morphology characteristic of growth from saline groundwater. This is consistent with their being potentially of seawater origin. The spatial distribution of the carbonate minerals may provide an indicator of domains where groundwater movement has been stagnant and differentiate these from domains where the flow system in this sedimentary rock sequence has been relatively dynamic.

Fossil shells occur at depths greater than approximately 50∼300 mbgl in the sedimentary strata (Fig. 2). The $\delta^{13}C_{PDB}$ of shell fossils are 0 ± 3‰. This suggests that, after sedimentation, the calcite has not been remobilized or recrystallised and the chemical condition of the deep groundwaters have remained saturated to super-saturated with respect to calcite. The depth of shell fossil occurrence roughly coincides with the depths of groundwater with a salinity between 5–10 g/L Total Dissolved Solid (Fig. 2). Thus the distribution of preserved fossil shells probably illustrates the limit of long-term infiltration and penetration of the dilute near-surface meteoric groundwater into the sedimentary sequence.

The $\delta^{13}C_{PDB}$ and $\delta^{18}O_{SMOW}$ of carbonate veins and nodules are approx. $+7\sim+21$‰ and $+24\sim+37$‰, respectively (Fig. 3). The $\delta^{18}O_{SMOW}$ data suggest that the calcite had precipitated at temperatures below 70°C. As shown by the preservation of shell fossils, groundwater flow is inferred to have been under long-term stagnant conditions in the deeper part of the sedimentary sequence. In contrast to the Tono area, there is a less clear depth relationship in the isotope data for carbonate minerals. The $\delta^{13}C_{PDB}$ values are very different from that of Tono and also that of the fossil shells. This possibly reflects the cycling of carbon during burial diagenesis in a closed system under static groundwater conditions in the Horonobe sedimentary rocks.

The concentration of dissolved inorganic carbon (DIC) ranges from 50 to 580 mg/L. The $\delta^{13}C_{PDB}$ values for DIC show an increase from approx. -9 to $+34‰$ with increasing depth, which is related to concentration (Fig. 5). The $\delta^{13}C_{PDB}$ value of DIC in ground-water is generally light (i.e. it shows negative values) as a result of isotopic fractionation during chemical/biochemical reaction among soil CO_2 ($\delta^{13}C_{PDB} \sim -25‰$), organic matter ($\delta^{13}C_{PDB}$ $-20 \sim -25‰$) and carbonate minerals in rock ($\delta^{13}C_{PDB}$; e.g. shell fossil $\sim 0‰$). The $\delta^{13}C_{PDB}$ value of DIC from shallow depths would be largely derived from these origins. Additionally, the generation of CH_4 and CO_2 from organic matter during organic matter maturation and biochemical reactions (Amo et al., 2007) is an important process that may influence the production of positive $\delta^{13}C_{PDB}$ values for DIC in the Horonobe area. Dissolved gas in the groundwater mainly consists of CH_4 and CO_2. The $\delta^{13}C_{PDB}$ value of CH_4 is about $-30‰$ at depths greater than 800 mbgl and $-41 \sim -56‰$ at depths shallower than 500 mbgl (JNOC, 1995; Ishiyama et al., 2008). These isotopic compositions for CH_4 are interpreted to indicate the formation of thermogenic CH_4 in the deeper part of the sequence, and to represent a mixture of biogenic CH_4 and thermogenic CH_4 in the relatively shallow part of the system. Ishiyama et al. (2008) show that $\delta^{13}C_{PDB}$ values of CO_2 are between $-3 \sim +12‰$ at depths above 500 mbgl. Biogenic methane gas is derived by decomposition of acetate, methanol and reaction of CO_2 and H_2. During these biochemical fermentation and reduction reactions, isotopically heavy CO_2 (with $\delta^{13}C_{PDB}$ of several tens ‰) is considered to be formed by isotopic enrichment through Rayleigh fractionation (Nissenbaum et al., 1972). The isotopic fractionation between dissolved bicarbonate ions and CO_2 gas is of the order of several ‰ at $0 - 70°C$ (Mook et al., 1973; Szaran, 1998). The heavy $\delta^{13}C_{PDB}$ values of DIC and calcite are probably the result of isotopic fractionation between organic matter, CO_2, DIC and calcite at various temperatures during the organic

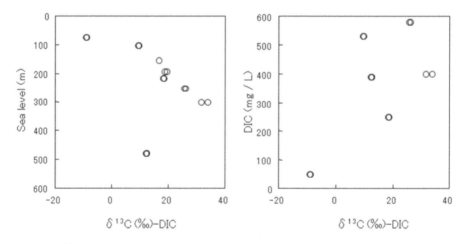

Figure 5. $\delta^{13}C_{PDB}$ values of DIC in groundwater in the Horonobe area.

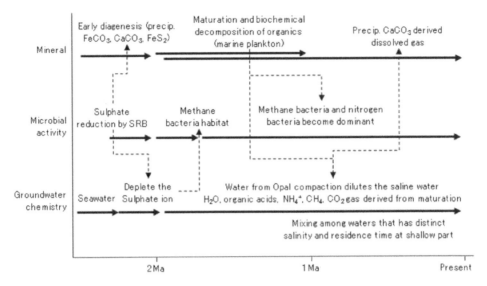

Figure 6.   Long-term hydrogeochemical processes in a deep closed system in the Horonobe area.

maturation process. Such interactions among water, mineral, microbe and gas are illustrated in Fig. 6. The sulphate ion reduction by sulphate reducing bacteria and precipitation of carbonate minerals would have occurred in initial diagenesis. Subsequent long-term biochemical maturation of organic matter provided the carbon dioxide into groundwater. The isotopically heavy calcite would be precipitated via such biochemical processes. The chemical and isotopic signature of carbonate minerals probably reflects the origin and bio-, geochemical processes in each age.

The isotopic data of calcite at Horonobe would appear to illustrate carbon cycling between groundwater, rock and gases in a closed system, rather than provide any evidence for the replacement of groundwaters along flow paths. This is because the groundwater system at depth has remained essentially under stagnant conditions since sedimentation, and has been unaffected by any flushing of the connate seawater by fresh water. Whereas in the shallow near-surface part of the sequence, there is a possibility that a different type of calcite has been precipitated in the transient zone between fresh water to saline water.

3.3   *Approach for crystalline and sedimentary rock area*

The mineralogical and geochemical information from secondary minerals for palaeo-hydrogeological analysis has previously been applied in studies of deep groundwater systems in crystalline rock in Sweden (Tullburg et al., 2007), and in sedimentary and crystalline metavolcanic rocks in the UK (Metcalfe et al. 1998; Milodowski et al., 1998). In these sites and in the Tono area, long-term chemical changes of groundwater can be estimated by evaluating the crystal growth morphology and isotopic/chemical records in secondary minerals such as calcite. Taking the static groundwater in a relatively young geological system such as in the Horonobe area into consideration, the use of hydrochemical studies to support hydrogeological investigations of the long-term behaviour of a groundwater system can be summarized as shown in Fig. 7.

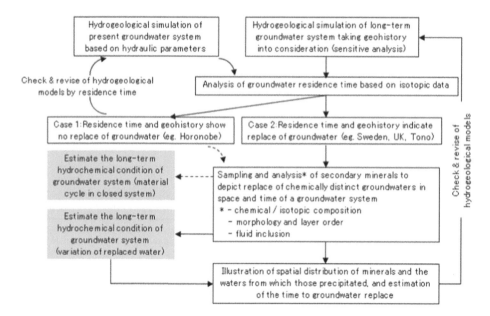

Figure 7. Hydrochemical approaches to hydrogeological study.

In relatively young geological systems inferred from considering the geohistory, hydrochemical evaluation based on groundwater residence time can be used to verify hydrogeological model simulations of groundwater system. In the closed system case at Horonobe, the mineralogical signature seems to reflect the local cycling of geochemical components between water, rock and diagenetically produced methane and carbon dioxide gases.

Whereas, in the groundwater system where the replacement of chemically distinct groundwaters occurred by subsequent groundwater flushing, the morphology and chemical/isotopic records of minerals would provide some constraints on groundwater retention in response to changes in recharge as a result of regional geomorphological evolution.

4 CONCLUSION

Detailed analyses of the spatial distribution of carbonate minerals, the occurrence (morphology, multilayered precipitation), and the isotopic/chemical signatures provide useful information to evaluate the long-term change or stability of groundwater systems in crystalline and sedimentary rocks. A crucial point of this approach is to identify evidence that might indicate that the replacement of chemically distinct groundwaters has occurred in the research areas under consideration. In any case, the observations from carbonate minerals precipitated contemporaneously within the modern groundwater system have the potential to be used to help place constraints on groundwater flow used in hydrogeological simulation studies.

ACKNOWLEDGEMENTS

The authors gratefully acknowledge for the staff members of Mizunami Underground Laboratory and Horonobe Underground Research Center. We also thank Tony Milodowski and Ian Holman for their help in preparing this manuscript.

REFERENCES

Amo, M., Suzuki, N., Shinoda, T., Ratnayake, N.P., Takahashi, K. (2007) Diagenesis and distribution of sterens in Late Miocene to Pliocene marine siliceous rocks from Horonobe (Hokkaido, Japan). Organic Geochemistry, Vol. 38: 1132–1145.

Chapman, N.A., McEwen, T.J. (1993) The application of paleohydrogeological information to repository performance assessment. In: Paleohydrogeological Methods and their Applications. Proc. NEA Workshop, Paris, 9–10 Nov. 1992, OECD, Paris, 117–146.

Emrich, K., Enhalt, D.H., Vogel, J.C. (1970) Carbon isotope fractionation during the precipitation of calcium carbonate. Earth and Planetary Science Letters, Vol. 8: 363–371.

Furue, R., Iwatsuki, T., Mizuno, T., Mie, H. (2003) Data book on groundwater chemistry in the Tono area. JNC TN 7450 2003-001. Japan Nuclear Cycle Development Institute, Japan.

Hama, K., Kunimaru, T., Metcalfe, R., Martin, A.J. (2007) The hydogeochemistry of argillaceous rock formations at the Horonobe URL site, Japan. Physics and Chemistry of the Earth, Vol. 32: 170–180.

Isihii, E., Yasue, K., Tanaka, T., Tsukui, R., Matsuo, K., Sugiyama, K., Matsuo, S. (2006) Three-dimensional distribution and hydrogeological properties of the Omagari Fault in the Horonobe area, northern Hokkaido, Japan. Jour. Geol. Soc. Japan, Vol. 112: 301–314.

Ishiyama, H., Watanabe, K., Waseda, A., Kato, S., Kunimaru, T. (2008) Study on the natural gas flow in Tertiary sedimentary rock in the Horonobe district based on the isotopic data. Paper presented at Japan Society of Civil Engineers 2008 Annual Meeting, Sendai, Japan, 10–12 September 2008.

Iwatsuki, T., Furue, R., Mie, H., Ioka, S., Mizuno, T. (2005) Hydrochemical baseline condition of groundwater at the Mizunami underground research laboratory (MIU). Applied Geochemistry, Vol. 20: 2283–2302.

Japan National Oil Corporation (JNOC) (1995) Report on the MITI Tempoku drilling (in Japanese).

Kanai, Y. (2007) Study on sorption of radioactive nuclides by carbonates-present and future researches-. Chikyukagaku (Geochemistry), Vol. 41: 1–16.

Metcalfe, R., Hooker, P.J., Darling, W.G., Milodowski, A.E. (1998) Dating Quaternary groundwater flow events: a review of available methods and their application. In: Parnell, J. (Ed.) 1998. Dating and Duration of Fluid Flow and Fluid-Rock Interaction. Geological Society, London, Special Publications, 144: 233–260.

Metcalfe, R., Hama, K., Amano, K., Iwatsuki, T., Saegusa, H. (2003) Geochemical approaches to understanding a deep groundwater flow system in the Tono area, Gifu-ken, Japan. Kono, In: Kono, I., Nishigaki, M. & Komatsu, M. (Eds), Groundwater Engineering—recent advances. A.A. Balkema, Tokyo, 555–561.

Milodowski, A.E., Gillespie, M.R., Naden, J., Fortey, N.J., Shepherd, T.J., Pearce, J.M., Metcalfe, R. (1998) The petrology and paragenesis of fracture mineralisation in the Sellafield area, West Cumbria. Procs. Yorks. Geol. Soc. 52: 215–241.

Mizuno, T., Iwatsuki, T. (2005) Study on long-term stability of geochemical environments at deep underground. Paper presented at the 15th Symposium on Geo-Environments and Geo-Technics, Yokohama, Japan, 10 December 2005.

Mook, W.G., Bommerson, J.C., Staverman, W.H. (1973) Carbon isotope fractionation between dissolved bicarbonate and gaseous carbon dioxide. Earth and Planetary Science Letters, Vol. 22: 169–176.

Niizato, T., Yasue, K., Kurikami, H., Kawamura, M., Ohi, T. (2008) Synthesizing geoscientific data into a site model for performance assessment; A study on the long-term evolution of the geological environmnet in and around the Horonobe URL, Hokkaido, northern Japan. Paper presented at the 3rd AMIGO Workshop on Approarches and Challenges for the use if geological information in the safety case, Nancy, France, 15–18 April 2008.

Niizato, T., Yasue, K. (2004) The long-term stability of the geological environments in and around the Horonobe area-Consideration of site specific features-. Paper presented at the 14th Symp. Geo-Environments and Geo-Technics, 101–106.

Nissenbaum, A., Preley, B.J., Kaplan, I.R. (1971) Early diagenesis in a reducing fjord,. Saanich Inlet, British Columbia. I. Chemical and isotopic changes in major components of. interstitial water. Geochim. Cosmochim. Acta, Vol. 36: 10915–10933.

Sasao, E., Ota, K., Iwatsuki, T., Niizato, T., Arthur, R.A., Stenhouse, M.J., Zhou, W., Metcalfe, R., Takase, H., Mackenzie, A.B. (2006) An overview of a natural analogue study of the Tono Uranium Deposit, central Japan. Geochemistry: Exploration, Environment, Analysis, Vol. 6: 5–12.

Szaran, J. (1998) Carbon isotope fractionation between dissolved and gaseous carbon dioxide. Chemical Geology, Vol. 150: 331–337.

Teramoto, M., Shimada, J., Kunimaru, T. (2006) Evidences of groundwater regime in impermeable rocks by stable isotopes in porewaters of drilled cores. Jour. Japan Soc. Eng. Geol., Vol. 47: 68–76.

Tullborg, E.-L., Drake, H., Sandstrom, B. (2008) Palaeohydrogeology: A methodology based on fracture mineral studies. Applied Geochemistry, Vol. 23: 1881–1897.

CHAPTER 9

Verification of ^{4}He and ^{36}Cl dating of very old groundwater in the Great Artesian Basin, Australia

Takuma Hasegawa
Central Research Institute of Electric Power Industry, Abiko-shi, Chiba-ken, Japan

Yasunori Mahara
Kyoto University. Kumatori-cho, Sennan-gun, Osaka, Japan

Kotaro Nakata
Central Research Institute of Electric Power Industry, Abiko-shi, Chiba-ken, Japan

M.A. Habermehl
Bureau of Rural Sciences, Canberra A.C.T., Australia

ABSTRACT: Groundwater dating is one of the most promising methods to evaluate very slow groundwater flow. In this study, ^{36}Cl and ^{4}He dating were carried out in the Great Artesian Basin, Australia. Groundwater samples were collected from 77 flowing artesian waterbores tapping aquifers in the Cadna-owie Formation and Hooray Sandstone, the uppermost confined aquifers in the Lower Cretaceous-Jurassic sedimentary sequence. Major ions, stable isotopes (δD, δ^{18}O and δ^{13}C), radio-active isotopes (^{14}C, ^{36}Cl) and noble gases (^{4}He, ^{3}He/^{4}He, Ne) were analysed. The results show that ^{36}Cl/Cl ratios decrease from the recharge areas to the discharge areas and ^{4}He values increase with the distances from the recharge areas. The ^{36}Cl ages have a close relationship with the distances from the recharge areas, as well as those from earlier studies. These relationships of ^{36}Cl concentrations reflect the groundwater residence times. Concentrations of ^{4}He increase with ^{36}Cl ages near the recharge areas. This result indicates that ^{4}He concentrations also reflect the groundwater residence times. However, ^{4}He concentrations vary due to sampling conditions, such as depressurisation and high temperatures in waterbores. These results confirm that ^{36}Cl and ^{4}He are useful indicators of groundwater residence times of very old groundwater.

Keywords: Groundwater dating, residence time, helium-4, chlorine-36, Great Artesian Basin

1 INTRODUCTION

Groundwater dating using radioisotopes is one of the most promising methods to assess very slow groundwater flow. The residence time of groundwater is important not only for groundwater use but also for contaminant transport. A number of groundwater dating methods are listed in Fig. 1. The applications of dating methods for groundwater older than ^{14}C dating time ranges are limited. In particular, if it is necessary to evaluate very long residence times for waste disposal of high-level radionuclides. In such cases ^{36}Cl and ^{4}He dating will be useful tools for safety assessments of waste disposal. Therefore, it is

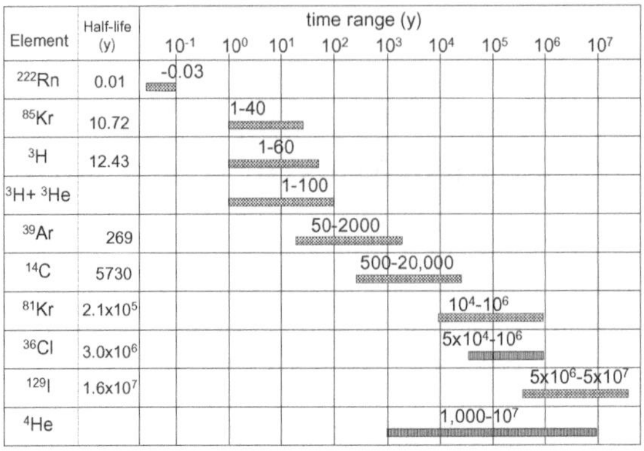

Element	Half-life (y)	time range (y)
^{222}Rn	0.01	-0.03 (10^{-1})
^{85}Kr	10.72	1-40 (10^0–10^1)
^{3}H	12.43	1-60 (10^0–10^1)
^{3}H+ ^{3}He		1-100 (10^1–10^2)
^{39}Ar	269	50-2000 (10^1–10^3)
^{14}C	5730	500-20,000 (10^2–10^4)
^{81}Kr	2.1×10^5	10^4–10^6
^{36}Cl	3.0×10^6	5×10^4–10^6
^{129}I	1.6×10^7	5×10^6–5×10^7
^{4}He		1,000-10^7

Figure 1.   Dating methods and applicable time ranges.

important to conduct and validate in-situ ^{36}Cl and ^{4}He dating investigations and such a study of ^{36}Cl and ^{4}He dating was conducted in the Great Artesian Basin, Australia.

2   PRINCIPLES OF ^{36}Cl AND ^{4}He DATING

2.1   *^{36}Cl dating*

Chloride is a relatively conservative component in the groundwater system and it is used as a tracer of groundwater flow. The chlorine radioisotope ^{36}Cl has a half-life of 3.01×10^6 y and can be used to characterize groundwater residence times. ^{36}Cl is produced by 1) cosmic ray interaction with ^{40}Ar and ^{36}Ar in the atmosphere and 2) in-situ thermal neutron activation of ^{35}Cl. The dating method using ^{36}Cl requires an initial input value in the recharge of groundwater and the measured ^{36}Cl/Cl ratio at the sampling point or the values of two or more sampling points along a groundwater flow line. The initial value of ^{36}Cl depends on the latitude and distance from the coast, as the strength of cosmic rays relates to the latitude and ^{36}Cl/Cl ratio is diluted by sea spray. The final ^{36}Cl/Cl ratio at the sampling point depends on the in-situ thermal neutron activation of ^{35}Cl, which relates to elements in the aquifer rock, i.e. uranium, thorium and other actinides.

Bentley et al. (1986) proposed the following equations for ^{36}Cl dating.

Equation (1) assumes that no sources or sinks exist in the groundwater system:

$$t = -\frac{1}{\lambda} \ln \left(\frac{R - R_{se}}{R_0 - R_{se}} \right) \tag{1}$$

where t is the age (y), λ ($=\ln(2)/T$) is the decay constant of ^{36}Cl, T is the half-life of ^{36}Cl (y), R is the measured ^{36}Cl/Cl ratio, R_0 is the initial ^{36}Cl/Cl ratio and R_{se} is the secular equilibrium ^{36}Cl/Cl ratio.

Equation (2) assumes that addition of chlorine has occurred and that the ^{36}Cl/Cl ratio is in secular equilibrium:

$$t = -\frac{1}{\lambda} \ln \left(\frac{C \, (R - R_{se})}{C_0 \, (R_0 - R_{se})} \right) \tag{2}$$

where C and C_0 are the measured and initial ^{36}Cl concentrations, respectively.

Equation (3) assumes the addition of dead chlorine, derived from rocks containing very old chloride such as evaporates, with a ^{36}Cl/Cl ratio of near zero.

$$t = -\frac{1}{\lambda} \ln \left(\frac{^{36}C - {}^{36}C_{se}}{^{36}C_0 - {}^{36}C_{se}} \right) \tag{3}$$

where ^{36}C, $^{36}C_0$, $^{36}C_{se}$ are measured, initial and secular equilibrium of ^{36}Cl concentrations.

2.2 ^{4}He dating

Helium is a noble gas and is inert and conservative. The helium isotope ^{4}He accumulates with time by in-situ production and external flux, with higher ^{4}He concentration indicating a longer residence time. ^{4}He is produced by the decay series of uranium and thorium that have very long half-lives and produce ^{4}He at a constant rate. An external flux of ^{4}He inflow occurs from outside the system. The equation of ^{4}He dating is expressed by the following equation:

$$t = \frac{D - D_{eq}}{M + F/nb} \tag{4}$$

where D is the measured helium concentration (cc_{STP}/g_w), D_{eq} is the atmospheric equilibrium concentration, M is the production rate from the rock ($cc_{STP}/g_w y$), n is porosity(-), b is aquifer thickness (cm) and F is the site dependent external flux ($cc_{STP}/cm^2 y$).

M is given by following equation.

$$M = \frac{(12.1 \cdot U + 2.9 \cdot Th)(1 - n)\, \rho_r\, (1 - L)}{1.0 \times 10^{14} \cdot n \cdot \rho_w} \tag{5}$$

where U is the uranium content (ppm), Th is the thorium content (ppm), n is porosity, ρ_r is rock density (Mg/m^3), ρ_w is water density (Mg/m^3) and L is the helium retention factor in the rock, which is usually negligible.

The external flux depends on the site because it relates to conditions of the surrounding environment. Mahara et al. (2007) proposed an external flux estimation based on the relationship between ^{4}He concentration and other dating results.

3 STUDY SITE

The Great Artesian Basin is one of the largest confined groundwater basins in the world. It occupies 1.7×10^6 km^2, about one-fifth of Australia (Fig. 2). The basin is bounded by the Great Dividing Range in the east and the high regions of central Australia to the west. The basin has a relatively simple bowl-like structure and contains a multi-layered system of quartz sandstone aquifers and intervening aquitards of siltstones and mudstones. This sedimentary sequence is exposed along the margins of the basin and tilted to the south-west. Habermehl (1980, 2001) provides more detailed hydrogeological information on the Great Artesian Basin.

The main groundwater recharge areas are along the basin's eastern margin on the western slopes of the Great Dividing Range and the dominant groundwater flow is directed in

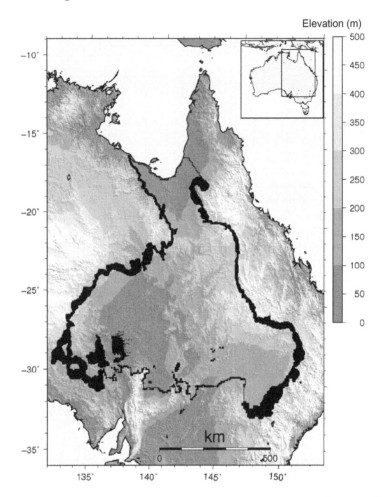

Figure 2.   Location of the Great Artesian Basin, Australia. The outcrop of the aquifers in the Cadna-owie Formation and Hooray Sandstone, the uppermost confined aquifers in the Lower Cretaceous-Jurassic sedimentary sequence are shown.

a southwest direction towards the southwestern marginal discharge areas as determined by previous studies and shown in Fig. 3 (Habermehl, 1980, 2001; Radke et al., 2000).

This study of very old groundwater dating in the Great Artesian Basin has been carried out on the aquifers in the Cadna-owie Formation and Hooray Sandstone, the uppermost confined aquifers in the Lower Cretaceous-Jurassic sedimentary sequence. The Great Artesian Basin has the advantages of:

1. Relatively simple groundwater flow because of simple geological structure;
2. Groundwater recharge within the elevated eastern outcrop areas and discharge at low elevations in the southwestern and southern discharge areas;
3. Groundwater inflow from outside the basin and mixing are limited;
4. Groundwater residence times are more than one million years because of very long flow lines and the hydrogeological and hydraulic characteristics of the aquifers;

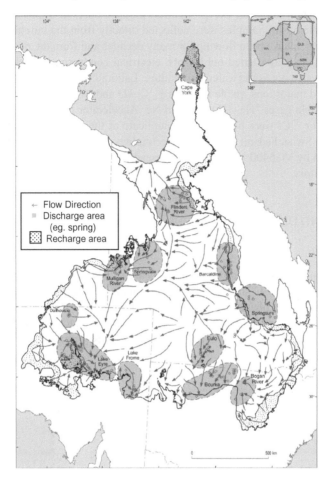

Figure 3. Regional artesian groundwater flow directions in the Cadna-owie Formation and Hooray Sandstone aquifers and their recharge and discharge areas, including locations of artesian springs.

5. No major tectonic events influenced the basin during the last few million years;
6. Many waterbores provide detailed and long term data and previous studies supply hydrogeological, hydrochemistry and isotope hydrology information.

4   WATER SAMPLING AND ANALYSIS

Groundwater samples were collected from 77 flowing artesian waterbores tapping the aquifers in the Cadna-owie Formation and Hooray Sandstone during 2002 and 2003. These aquifers are the main aquifers exploited in the basin and were also sampled for previous studies by Bentley et al. (1986), Torgersen et al. (1985), Torgersen et al. (1991), Torgersen et al. (1992), Love et al. (2000) and Lehmann et al. (2003). The boreholes sampled were selected on the basis of the groundwater flow paths as shown in Fig. 3. Most of the bores are flowing artesian waterbores, except for a few pumped bores near the eastern and western margins of the basin. Samples were collected directly from the boreheads of flowing artesian bores which have been flowing for many decades and from the pumped bores. In-field measurements were carried out for pH, electrical conductivity, oxidation-reduction potential, dissolved oxygen and water temperature. Laboratory analyses of the groundwater samples included major ions, stable isotopes $\delta^{13}C$, δD and $\delta^{18}O$, radioactive isotopes ^3H, ^{14}C and ^{36}Cl, noble gases ^4He, $^3He/^4He$ and Ne. Accelerator mass spectrometry (AMS) measurements for ^{36}Cl were performed in duplicate at the Australian National University (ANU) and Swiss Federal Institute of Technology (ETH). Noble gas measurements were undertaken by VG5400 in the Central Research Institute of Electric Power Industry (CRIEPI) laboratory.

5   RESULT AND DISCUSSION

5.1   *Main results*

The laboratory analysis results of the major ions are shown as meq/L in Stiff diagrams at the bore locations in Fig. 4. The main regional groundwater flow paths are also shown in Fig. 4. A general trend of increasing Total Dissolved Solid (TDS) values exists along most of the groundwater flow paths. Geological features, such as faults and shallow basement areas, ridges or arcs influence the hydrogeology, groundwater flow paths and hydrochemistry.

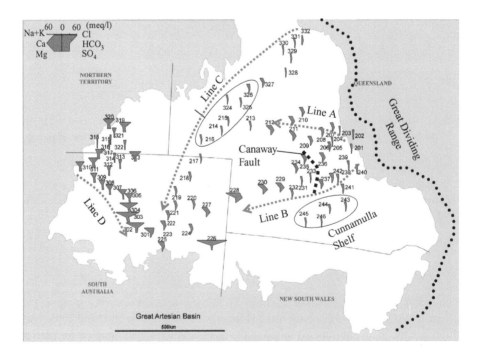

Figure 4.   Stiff diagrams (major ions) shown at the locations of waterbores sampled and regional groundwater flow paths.

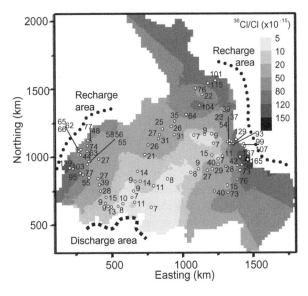

Figure 5. ^{36}Cl/Cl distribution, including recharge and discharge areas.

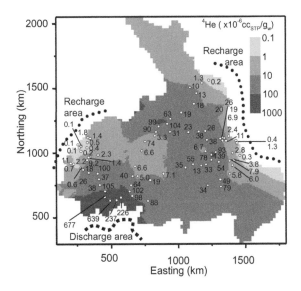

Figure 6. ^{4}He distribution, including recharge and discharge areas.

The distribution of ^{36}Cl/Cl ratios is shown in Fig. 5 and was drawn using the kriging geostatistical estimation method to correlate the measured data. The ^{36}Cl/Cl ratios decrease with distance from the recharge areas due to radioactive decay and this trend is evident up to many hundreds of kilometres from the recharge areas.

The ^{4}He distribution is shown in Fig. 6 and was also drawn by kriging. The ^{4}He concentrations increase with distance from the recharge areas because of ^{4}He accumulation within the artesian groundwater. However, samples from the centre of the basin show rather low

[4]He concentrations. The artesian groundwater in those areas is under high pressure at high temperature because of the depth of the aquifers at thousands of metres below the ground surface and high geothermal gradients. Artesian groundwater discharges from some bores at 100°C and dissolved [4]He may be degassing by depressurisation in the groundwater flow to the surface in the bores. Significant accumulations of [4]He occur within hundreds of kilometres from the recharge areas.

5.2   Discussion on [36]Cl dating

In previous studies, [36]Cl/Cl ratios were plotted against [36]Cl to understand the effect of dilution, evaporation and addition of dead chlorine. Similar plots for [36]Cl/Cl ratios determined in this study along each regional groundwater flow path are shown in Fig. 7. Line A shows [36]Cl/Cl ratios which plot approximately on the line that expresses radioactive decay. The influence of evaporation, dilution, addition of dead chlorine and mixing of groundwater of different origin appears to be negligible at Line A. At Line B, the data from the eastern recharge areas towards the southwest across the Canaway Fault are also on or close to the [36]Cl/Cl decay line with some variations. The data from bores beyond the Canaway Fault and near the Cunnamulla Shelf are not on the same line. Habermehl (1986) showed that upward vertical leakage takes place from one aquifer to another along the fault zone. At

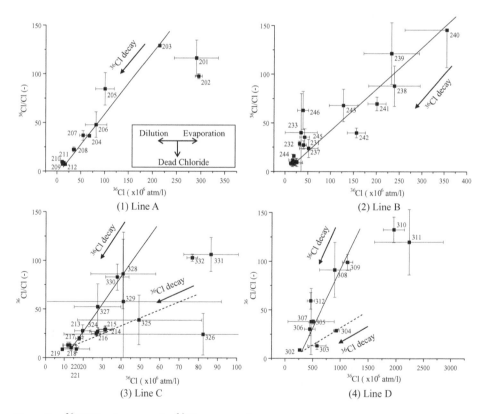

Figure 7.   [36]Cl/Cl ratio versus the [36]Cl concentration.

Line C, it appears that two trends are present with different relationships to the ^{36}Cl/Cl decay line as well as some data with large errors. A comparison of Figs. 3 and 4 show that Line C is a composite flow path, with ^{36}Cl and ^{36}Cl/Cl decreasing at a constant rate from the northeastern recharge areas to the Queensland-South Australian border and then to the southwestern basin margin. Mixing with relatively modern water occurs in the central part of the flowline. Torgersen et al. (1991) also pointed to the possibility of mixing. At line D, two trends were identified and there are similarities to Line C with probable mixing. The data on the dashed line show bores with higher total dissolved solids from water of a different origin than the other bores along the flowline shown in Fig. 4.

Fig. 8 shows plots of the ^{36}Cl ages against the distances from recharge areas for each flow path. The ^{36}Cl ages were calculated using the three equations listed in section 2.1. The relationship between ^{36}Cl ages and distances demonstrate a close relationship and ^{36}Cl ages are proportional to the distances. Close relationships are apparent at Lines A and B. The data from bores located at Line B beyond the Canaway Fault and near the Cunnamulla Shelf were excluded, because of the mixing of different groundwater as shown in Fig. 7. Line C in Fig. 8 clearly shows two separate trends, that display the relationships between ^{36}Cl ages and distances. As shown above, this is caused by the mixing of groundwater from different flowpaths. The mixing with modern water occurred in a limited area and after the mixing ^{36}Cl ages increase again with distance. At Line D, dissolved major ions vary as shown in Fig. 4. Therefore, the mixing of groundwater from different origins occurs at the flow paths as expressed by the dashed and full lines in Fig. 7(4).

The ^{36}Cl ages obviously increase along the flow paths. The calculated velocities based on this relationship between ^{36}Cl age and distance are shown in Fig. 8. The range of velocity

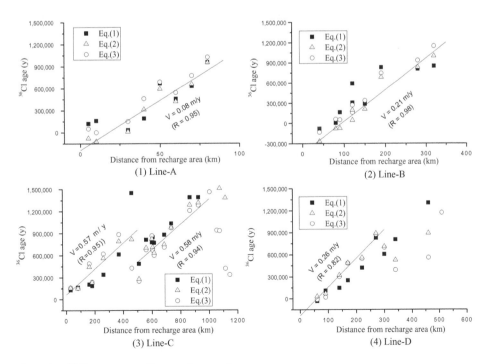

Figure 8. ^{36}Cl age versus distance from recharge area.

is from 0.08–0.54 m/y. These values are in good agreement with hydrological information. These results indicate that ^{36}Cl age express the residence time by considering the flow path.

5.3 Discussion of ^{4}He dating

^{4}He ages are calculated by dividing the ^{4}He excess concentration values by the accumulation rate. ^{4}He accumulation rates are difficult to estimate because the external flux is unknown. We use the ^{4}He concentration instead of ^{4}He ages and plot ^{4}He concentrations against distances from recharge area as shown in Fig. 9. The logarithm of the ^{4}He concentrations were plotted because ^{4}He concentrations have large variations. Possible causes might be degassing due to high temperatures and depressurisation. Plots of ^{3}He/^{4}He and Ne against distance are also shown in Fig. 9. ^{3}He/^{4}He is useful to identify the origin of helium, including radiogenic, mantle and atmospheric. The characteristic ^{3}He/^{4}He ratio of radiogenic, mantle and atmospheric sources are 1.0×10^{-8}, 1.0×10^{-5} and 1.4×10^{-6}, respectively and therefore, ^{3}He/^{4}He will be close to 1.0×10^{-8} when radiogenic helium accumulates.

Measurement of ^{3}He/^{4}He from the bores sampled show that most helium is radiogenic in origin. Ne is one of the contamination indicators of air. Atmospheric equilibrium of Ne is 2.0×10^{-7} cc_{STP}/g_w and if air contamination occurs, Ne concentrations will be high. Measurements of Ne concentrations from the bores sampled indicate higher than atmospheric equilibrium. As Ne is almost constant in the Great Artesian Basin bores sampled, air contamination probably did not occur during the sampling.

Concentrations of ^{4}He increase with distances along all of the Lines A, B, C and D, and ^{4}He ages also increase with distances as with ^{36}Cl ages. At Line B, ^{4}He concentrations near the Canaway Fault are higher than elsewhere and a similar trend was observed for ^{36}Cl ages. Old groundwater from lower confined aquifers may flow upwards through the Canaway Fault zone. At Line C, the ^{4}He concentrations increase in two steps from the recharge areas to the centre of Line C and from there to the discharge area near the basin margin. These results confirm that mixing of modern groundwater occurs in the central part of the flow path. At Line D, some data show large variations, which might be caused by degassing, because some groundwater samples were collected by pumping the bores in the southwestern basin margin. Concentrations of ^{4}He in the centre of the basin are low and exhibit large variations. The latter are probably also caused by sampling conditions. The aquifer sampled in the central part of the basin is more than two thousands metres below the ground surface and the groundwater temperatures and water pressures are high. The possibility of degassing during sampling is high due to depressurisation and it is important to improve the sampling methods for these conditions. ^{4}He concentrations could reflect the groundwater residence times and therefore ^{4}He sampling requires special care so that ^{4}He concentrations represent actual values and do not show variations caused by degassing.

^{4}He concentrations were compared to ^{36}Cl ages to verify both results. ^{4}He concentrations were plotted against ^{36}Cl ages as shown in Fig. 10. The data from bores near the recharge areas were used and the data influenced by mixing were excluded. A regression analysis was conducted to evaluate the correlation and ^{4}He accumulation rates. At Line A, B and C there are good correlations between ^{4}He concentrations and ^{36}Cl ages and the ^{4}He accumulation rates are much higher than the in-situ production derived from Equation 5. It indicates that external fluxes are the dominant sources of ^{4}He accumulation. The differences in ^{4}He accumulation rates are influenced by geological conditions, such as aquifer

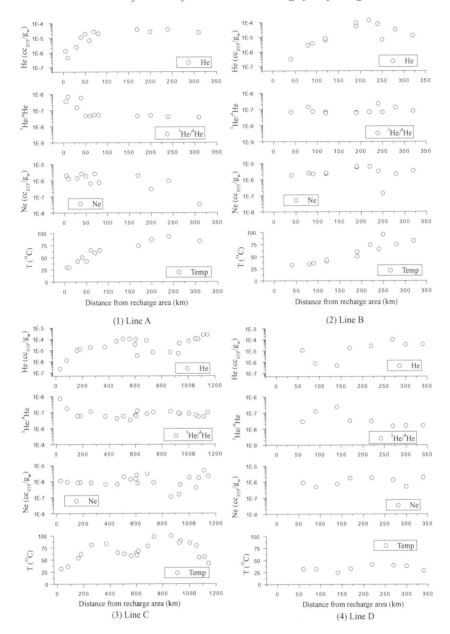

Figure 9.   Noble gases versus distance from recharge.

thickness, porosity and leakage rates from the aquifers and aquitards. The difference in accumulation rates are within approximately 3 times. At Line D the correlation is not good and ⁴He accumulation rates are much higher than along the other lines. Line D is located in the southwestern part of the basin, where some groundwater samples were collected by pumping. The external flux in the western part of the basin is larger than in most other parts of the basin, as the aquifer directly overlies igneous and metamorphic basement rocks.

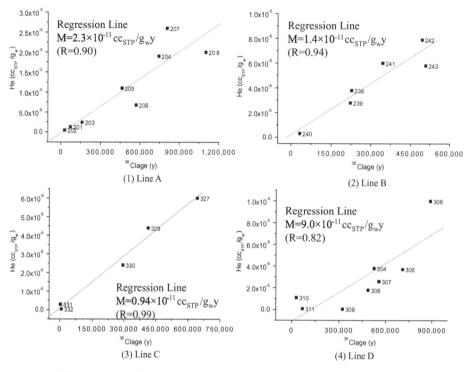

Figure 10. ^{36}Cl age versus ^{4}He concentration near recharge area.

In many other parts of the basin older sedimentary basins underlie the Great Artesian Basin and these rock systems reduce the upwards flux from deeper sources. It is apparent that ^{4}He concentrations reflect groundwater residence times as there is a clear correlation between ^{4}He concentrations and ^{36}Cl ages.

6   SUMMARY

We applied ^{36}Cl and ^{4}He dating to artesian groundwater in the Great Artesian Basin. ^{36}Cl ages along the defined flow paths are proportional to the distances up to hundreds of kilometres from the recharge areas as was also shown by several previous studies. Groundwater residence times are reflected by ^{36}Cl ages, and it is clear that ^{4}He concentrations also reflect groundwater residence times as ^{4}He concentrations increase with distances from the recharge areas and show a close relationship with ^{36}Cl ages. The validity of ^{36}Cl dating results were proven in several previous studies by comparing hydrologic ages, calculated from hydraulic conductivities, hydraulic gradients and porosity values. In this study, ^{36}Cl and ^{4}He dating results were compared with each other, although ^{4}He dating is constrained by difficulties in determining the external fluxes, which depend on site specific conditions. In this study ^{4}He accumulation rates were estimated from the relationship between ^{4}He concentrations and ^{36}Cl ages and this method will assist the determination of ^{4}He accumulation rates and the interpretation of ^{4}He dating. All of our results confirm that

^{36}Cl and ^{4}He dating methods are useful to identify and characterise very old groundwater. However, the evaluation of groundwater mixtures of different origin is difficult and the determination of groundwater ages in the aquifers at great depths in the centre of the Great Artesian Basin is a problem.

ACKNOWLEDGEMENTS

We thank Prof. A. Kudo (Kyoto Univ.), Prof. T. Nakamura (Nagoya Univ.), Prof. J. Shimada (Kumamoto Univ.), Prof. M. Nishigaki (Okayma Univ.) and Assoc. Prof. T. Tokunaga for their advice to this project. We also thank Y. Mizuochi, H. Kobayashi, A. Ninomiya (SUMICON) and T.R. Ransley (BRS) for assistance during sampling trips and to Prof. L.K. Fifield (Australian National Univ.) and Prof. M. Suter (Swiss Federal Institute of Technology) for ^{36}Cl AMS measurements. This study entitled 'Research and development on groundwater dating technique' was carried out under contracts awarded by the Japanese Ministry of Economy, Trade and Industry (METI) to CRIEPI.

REFERENCES

Bentley, H.W., Phillips, F.M., Davis, S.N., Habermehl, M.A., Airey, P.L., Calf, G.E., Elmore, D., Gove, H.E. and Torgersen, T. (1986) Chlorine 36 dating of very old groundwater: 1.The Great Artesian Basin, Australia, Water Resour. Res., 22: 1991–2001.

Habermehl, M.A. (1980) The Great Artesian Basin, Australia. BMR Journal of Australian Geology & Geophysics, 5: 9–38.

Habermehl, M.A., (1986) Regional groundwater movement, hydrochemistry and hydrocarbon migration in the Eromanga Basin. In: Graveostock, D.I., Moore, P.S. and Pitt, G.M. (Editors), Contributions to the geology and hydrocarbon potential of the Eromanga Basin. Geological Society of Australia Inc. Special Publication 12: 353–376.

Habermehl, M.A. (2001) Hydrogeology and Environmental Geology of the Great Artesian Basin, Australia, In: Gostin, V.A. (Editor), Gondwana to Greenhouse—Australian Environmental Geoscience. Geological Society of Australia Inc., Special Publication, 21: 127–143, 344–346.

Lehmann, B.E., Love, A., Purtschert, R., Collon, P., Loosli, H.H., Kutschera, W., Beyerle, U., Aeschbach-Hertig, W., Kipfer, R., Frape, S.K., Herczeg, A., Moran, J., Tolstikhin, I.N., and Gronin, M. (2003) A comparison of groundwater dating with ^{81}Kr, ^{36}Cl and ^{4}He in four wells of the Great Artesian Basin, Australia, Earth and planetary Science Letters, 211: 237–250.

Love, A.J., Herczeg, A.L., Sampson, L., Cresswell, R.G. and Fifield, L.K. (2000) Sources of chloride and implications for ^{36}Cl dating of old groundwater, southwestern Great Artesian Basin, Australia. Water Resour. Res., 36: 1561–1574.

Mahara, Y., Habermehl, M.A., Miyakawa, K. and Shimada, J. (2007) Can the ^{4}He clock be calibrated by ^{36}Cl for groundwater dating? Nuclear Instruments and Methods in Physics Research B, 259: 539–546.

Radke, B.M., Ferguson, J., Cresswell, R.G., Ransley, T.R. and Habermehl, M.A. (2000) The hydrochemistry and implied hydrodynamics of the Cadna-owie-Hooray Aquifer, Great Artesian Basin, Australia. Bureau of Rural Sciences, Canberra, 229 p.

Torgersen, T. and Clarke, B.W. (1985) Helium accumulation in groundwater I: An evaluation of sources and the continental flux of crustal ^{4}He in the Great Artesian Basin, Australia. Geochimica et Cosmochimica Acta, 49: 1211–1218.

Torgersen, T., Habermehl, M.A., Phillips, F.M., Elmore, D., Kubik, P., Jones, B.G., Hemmick, T. and Gove, H.E. (1991) Chlorine-36 dating of very old groundwater: 3.Further studies in the Great Artesian Basin, Australia, Water Resour. Res., 27: 3201–3213.

Torgersen, T., Habermehl, M.A. and Clarke, B.W. (1992) Crustal helium fluxes and heat flow in the Great Artesian Basin, Australia, Chemical Geology, 102: 139–152.

CHAPTER 10

Land subsidence characteristics of the Jakarta basin (Indonesia) and its relation with groundwater extraction and sea level rise

Hasanuddin Z. Abidin, Heri Andreas, M. Gamal & Irwan Gumilar
Geodesy Research Division, Institute of Technology Bandung, Bandung, Indonesia

Maurits Napitupulu
Provincial Industry and Energy Agency of DKI Jakarta, Indonesia

Yoichi Fukuda
Graduate School of Science, Kyoto University, Japan

T. Deguchi & Y. Maruyama
Earth Remote Sensing Data Analysis Centre (ERSDAC), Tokyo, Japan

Edi Riawan
Centre for Coastal and Marine Development, Institute of Technology Bandung, Bandung, Indonesia

ABSTRACT: Jakarta is the capital city of Indonesia with a population of about 9 million people, inhabiting an area of about 660 km^2. It has been reported for many years that several places in Jakarta are subsiding at different rates. Over the period of 1982–1997, subsidence ranging from 20 to 200 cm is evident in several places in Jakarta. There are four different types of land subsidence that can be expected to occur in the Jakarta basin, namely: subsidence due to groundwater extraction, subsidence induced by the load of constructions (i.e. settlement of high compressibility soil), subsidence caused by natural consolidation of alluvial soil and tectonic subsidence. In addition to the levelling surveys, GPS survey methods and InSAR measurements have been used to study land subsidence in Jakarta. This paper describes the characteristics of subsidence in Jakarta over the period of 1982 to 2007 as observed by the three methods. In general land subsidence in Jakarta exhibits spatial and temporal variations, with rates of about 1 to 15 cm/year. A few locations can have subsidence rates up to about 20–25 cm/year. It was found that the spatial and temporal variations in land subsidence correlate with variations in groundwater extraction, coupled with the characteristics of sedimentary layers and building loads above it. The observed subsidence rates in several locations show a positive correlation with known volumes of groundwater extraction. However, the relative magnitude and spatial variability of the effect of groundwater extraction on land subsidence in the whole Jakarta basin is not yet fully understood. In the coastal areas of Jakarta, the combined effects of land subsidence and sea level rise also introduce other collateral hazards, namely the tidal flooding phenomena.

Keywords: Jakarta, land subsidence, levelling, GPS, InSAR, groundwater, sea level rise

1   INTRODUCTION

Land subsidence is not a new phenomenon for Jakarta, the capital city of Indonesia. It has been reported for many years that several areas of Jakarta are subsiding at different rates (Murdohardono & Tirtomihardjo, 1993; Murdohardono & Sudarsono, 1998; Rajiyowiryono, 1999). The impact of land subsidence in Jakarta is observed in several forms, such as cracking of permanent constructions and roads, wider expansion of flooding areas, malfunction of drainage systems, and increased inland sea-water intrusion.

Based on several studies (Murdohardono and Sudarsono, 1998; Rismianto and Mak, 1993; Harsolumakso, 2001; Hutasoit, 2001), there are four different types of land subsidence that can be expected to occur in the Jakarta basin, namely: subsidence due to groundwater extraction, subsidence induced by the load of constructions (i.e. settlement of high compressibility soil), subsidence caused by natural consolidation of alluvial soil, and geotectonic subsidence. The first three are thought to be the dominant types of land subsidence in Jakarta basin.

In the case of Jakarta, comprehensive information on the characteristics of land subsidence is applicable to several important planning and mitigation efforts (see Fig. 1), such as spatial-based groundwater extraction regulation, effective control of flood and seawater intrusion, environmental conservation, design and construction of infrastructure, and spatial development planning. Considering the importance of land subsidence information for supporting development activities in the Jakarta area, monitoring and studying the characteristics of this subsidence phenomenon becomes more valuable.

Since the early 1980's, land subsidence in several areas of Jakarta has been measured using a variety of measurement techniques including levelling surveys, extensometer measurements, groundwater level observations, GPS (Global Positioning System) surveys, and InSAR (Interfero-metric Synthetic Aperture Radar) (Abidin, 2005; Abidin et al., 2001, 2004, 2008). The prediction of ground subsidence, based on models incorporating geological and hydrological parameters of Jakarta, has also been investigated (Murdohardono and Tirtomihardjo, 1993; Yong et al., 1995; Purnomo et al., 1999).

This paper describes the characteristics of land subsidence in the Jakarta basin during the period of 1982 to 2007, as observed by levelling surveys, GPS surveys and InSAR. Theoretically, groundwater extraction that exceeds aquifer recharge could lead to a decline in groundwater levels (piezometric head), reducing the hydrostatic pressures supporting

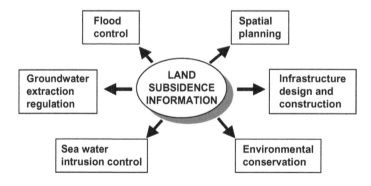

Figure 1.   The importance of land subsidence information.

the aquifer material and causing land subsidence. Therefore the correlation between the observed subsidence and groundwater extraction volumes in the Jakarta area will also be studied.

2 JAKARTA AND ITS CHARACTERISTICS

The city of Jakarta has a population of about 9 million people (BPS Jakarta, 2007), inhabiting an area of about 661.52 km^2. Jakarta is located on the lowland of the northern coast of the West Java province (centred at coordinates of about $-6°15'$ latitude and $+106°50'$ longitude), as shown in Fig. 2. The area is relatively flat, with topographical slopes ranging between 0° and 2° in the northern and central parts, and between 0° and 5° in the southern part. The southernmost area of Jakarta has an altitude of about 50 m above mean sea level (MSL). The average annual rainfall in the Jakarta area is about 2000 mm/year, with the maximum monthly average occurring in January and the minimum occurring in September.

Regionally speaking, Jakarta is a lowland area which has five main landforms, namely: (1) volcanic alluvial fan landforms, which are located in the southern part; (2) landforms of marine-origin, which are found in the northern part adjacent to the coastline; (3) beach ridge landforms, which are located in the northwest and northeast parts; (4) swamp and mangrove-swamp landforms, which are encountered in the coastal fringe; and (5) former river channels, which run perpendicular to the coastline (Rimbaman and Suparan, 1999; Sampurno, 2001). It should also be noted that there are about 13 natural and artificial rivers flowing through Jakarta, of which the main rivers, such as Ciliwung, Sunter, Pesanggrahan, Grogol and their tributaries, form the main drainage system of Jakarta.

In terms of geological and hydrological settings, according to Yong et al. (1995), the Jakarta basin consists of a 200 to 250 m thick sequence of Quaternary deposits which overlies Tertiary sediments. The base of the Quaternary deposits has been defined as the lower boundary of the groundwater basin. The Quaternary sequence can be further subdivided into three major units, which, in ascending order are: a sequence of Pleistocene marine and non-marine sediments, a late Pleistocene volcanic fan deposit, and Holocene marine and floodplain deposits.

Three aquifers are recognized within the 250 m thick sequence of Quaternary sediment of the Jakarta basin, namely: the Upper Aquifer, an unconfined aquifer, occurs at a depth

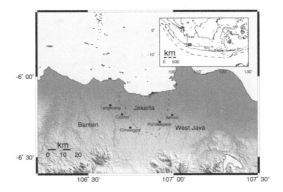

Figure 2.   Jakarta and its surrounding areas.

of less than 40 m; the Middle Aquifer, a confined aquifer, occurs at a depth between 40 and 140 m; and the Lower Aquifer, a confined aquifer, occurs at a depth between 140 and 250 m (Soetrisno et al.,1997; Hadipurwo,1999). The geologic materials confining these aquifers are silt and clay. Inside those aquifers, the groundwater generally flows from south to the north (Lubis et al., 2008). Below a depth of 250 m, an aquifer in the Tertiary sediments also has been identified. But according to Murdohardono and Tirtomihardjo (1993), it is less productive and its water quality is relatively poor.

3   LAND SUBSIDENCE AS OBSERVED BY LEVELLING SURVEYS

The first reliable information about subsidence in Jakarta came from results of levelling surveys. The systematic levelling surveys covering much of the Jakarta area were conducted in 1978, 1982, 1991, 1993, and 1997. Except for the last survey, which was performed by the Local Mines Agency of Jakarta, the levelling surveys were done by the Local Surveying and Mapping Agency of Jakarta. The levelling surveys were done using Wild N3, Zeiss Ni002, and Wild NAK precise levelling instruments. Each levelling line was measured in double-standing mode, and each levelling session was measured forward and backward to provide survey closure and to verify accuracy. The levelling line for each session is about 1 km in length. The tolerance for the difference between the forward and backward height-difference measurements is set to be $4\sqrt{D}$ mm, where D is the length of levelling line in km.

After applying relatively strict quality assurance criteria, only three surveys were considered sufficiently accurate for investigating the land subsidence in Jakarta; those conducted in 1982, 1991, and 1997. Moreover, only the results from specific levelling points in the network, which are considered the most reliable, are used for investigating land subsidence. In this case, repeatability of the heights obtained from different surveys and different loops, and stability of the monument with respect to its local environment, are used as the main criteria for selecting the points. The distribution of these levelling points is shown in Fig. 3.

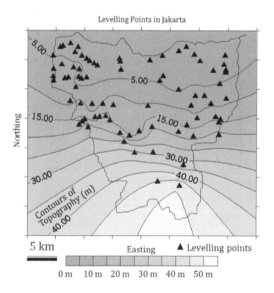

Figure 3.   Levelling network in Jakarta.

The network consists of about 80 points distributed across Jakarta. The magnitude of land subsidence was estimated using 45 selected points from the levelling networks of 1982, 1991, and 1997.

Based on those levelling surveys, over the 15-year period of 1982 to 1997, subsidence ranging from 20 cm to 200 cm is evident at several places in Jakarta. Fig. 4 shows that the maximum land subsidence observed by the levelling surveys during the period of 1991–97 is about 160 cm, while for the period of 1982–91 it is about 80 cm. The rates of land subsidence during the period of 1991–97 are also, in general, larger than those during the 1982–91 period, as indicated by the box-and-whisker plot shown in Fig. 5.

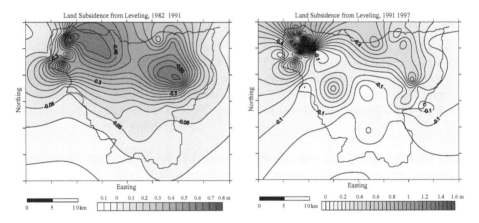

Figure 4. Land subsidence in Jakarta measured from levelling surveys (in metres), Over the periods of 1982–1991 (left) and 1991–1997 (right).

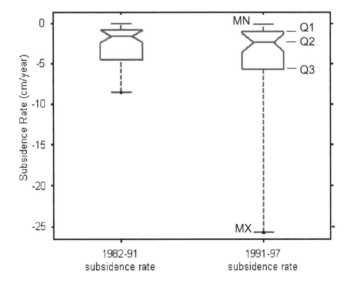

Figure 5. Box-and-Whisker plots of levelling-derived land subsidence rates in Jakarta. In this Figure: MN = sample minimum, Q1 = lower quartile (25th percentile), Q2 = median (50th percentile), Q3 = upper quartile (75th percentile), and MX = sample maximum.

In general the subsidence rates in Jakarta area during this period are about 1–5 cm/year and can reach 25 cm/year at several locations (see Fig. 5). From Fig. 4 it can be seen that land subsidence in the northern part of Jakarta, which is close to the sea, is larger than in the southern part of Jakarta. In this case, three regions, namely two in the northwestern part (Cengkareng and Kalideres districts) and one in the northeastern part of Jakarta (Kemayoran-Sunter district), show the largest subsidence compared to the other regions. More comprehensive results on levelling-based subsidence in Jakarta can be seen in Abidin et al. (2001).

4   LAND SUBSIDENCE AS OBSERVED BY GPS SURVEYS

Besides using levelling surveys, land subsidence in Jakarta has also been studied using GPS survey methods (Abidin et al., 2002; Leick, 2003; Abidin, 2007). The GPS-based land-subsidence study has been conducted by the Geodesy Research Division of ITB since December 1997. Up to now, ten GPS surveys has been conducted, namely on: 24–26 December 1997, 29–30 June 1999, 31 May–3 June 2000, 14–19 June 2001, 26–31 October 2001, 02–07 July 2002, 21–26 December 2002, 21–25 September 2005, 10–14 July 2006 and 3–7 September 2007.

The configuration of this GPS monitoring network is shown in Fig. 6. These surveys did not always occupy the same stations. The first survey started with 13 stations, and the network then expanded subsequently to 27 stations. At certain epochs, some stations could not be observed due to the destruction of monuments, or severe signal obstruction caused by growing trees and/or new construction. Station BAKO is the southernmost point in the

Figure 6.   GPS network for monitoring subsidence in Jakarta.

network and is also the Indonesian zero order geodetic point. It is used as the reference point. In this case the relative ellipsoidal heights of all stations are determined relative to BAKO station. BAKO is an IGS station, operated by the National Coordinating Agency for Survey and Mapping of Indonesia.

The GPS surveys exclusively used dual-frequency geodetic-type GPS receivers. The length of surveying sessions was in general between 9 to 11 hours. The data were collected with a 30 second interval using an elevation mask of 15°. The data were processed using the software Bernese 4.2 (Beutler et al., 2001). Since we are mostly interested with the relative heights with respect to a stable point, the radial processing mode was used instead of network adjustment mode. The standard deviations of GPS-derived relative ellipsoidal heights from all surveys were in general better than 1 cm (Abidin et al., 2008).

Examples of GPS-derived land subsidence at several observing stations are shown in Figs. 7, 8 and 9. On these Figures, the first measurement at each point established the baseline elevation for that site and, therefore, is shown as zero elevation change. The

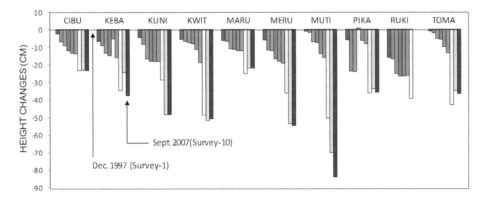

Figure 7.   Accumulated GPS derived subsidence (cm) during the period of Dec. 1997 to Sept. 2007. The baseline elevations are those from GPS Survey-1 (Dec. 1999).

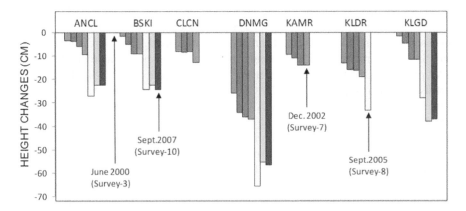

Figure 8.   Accumulated GPS derived subsidence (cm) during the period of June 2000 to Sept. 2007. The baseline elevations are those from GPS Survey-3 (June 2000).

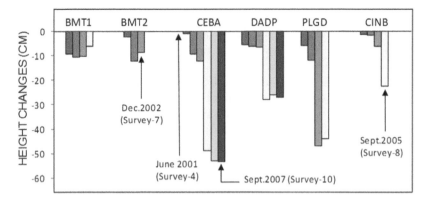

Figure 9.   Accumulated GPS derived subsidence (cm) during the period of June 2001 to Sept. 2007. The baseline elevations are those from GPS Survey-4 (June 2001).

different color codes for the individual bars in these figures indicate that they were estimated from different GPS surveys.

In about ten years, i.e. Dec. 1997 to Sept. 2007, the accumulated subsidence at several GPS stations can reach about 80–90 cm. In general the GPS observed subsidence rates during the period between December 1997 and September 2007 are about 1–15 cm/year. It is also found that land subsidence rates in the Jakarta basin have both a spatial and a temporal variation. This indicates that the causes of land subsidence in Jakarta also differ spatially. A more comprehensive review of GPS-derived land subsidence in Jakarta can be found in Abidin (2005) and Abidin et al. (2001, 2008).

5   LAND SUBSIDENCE AS OBSERVED BY INSAR

Since 2004, subsidence phenomena in Jakarta also have been studied using InSAR (*Interferometric Synthetic Aperture Radar*) techniques (Schreier, 1993; Massonnet and Feigl, 1998). The initial results are provided in Abidin et al. (2004). Recently, InSAR techniques were applied to study land subsidence in the Jakarta area using data from the ALOS/PALSAR satellite, which was launched in January 2006 as a successor of JERS-1/SAR. The SAR data observed during June 2006 and February 2007 were used. They were acquired in Fine Beam Single Polarization mode (HH polarization) with off-nadir angle of 41.5 degrees. InSAR processing has been performed using Level 1.1 products (SLC: Single Look Complex) distributed to the public by ERSDAC (Earth Remote Sensing Data Analysis Center) in Japan. The processing software for InSAR was developed by Deguchi (2005) and Deguchi et al. (2006).

Fig. 10 shows land subsidence in northern part of Jakarta detected by InSAR over the period of June 2006 to February 2007. In this Figure, subsidence is calculated by multiplying the number of colour fringes by 11.8 cm. This Figure shows that subsidence along the coastal zone of Jakarta has a spatial variation. After correlation with GPS results, it was found that the maximum InSAR-derived subsidence rates for the eight month period between InSAR measurements reached about 12 cm/year, as shown in Fig. 11. This subsidence rate is comparable with the rate observed by the GPS and levelling surveys.

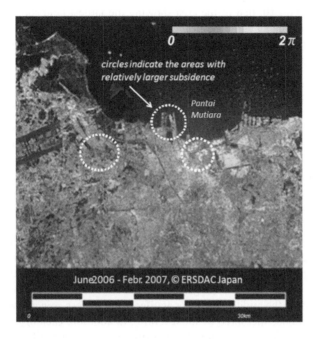

Figure 10.   InSAR-derived subsidence in the northern part of Jakarta using ALOS PALSAR data.

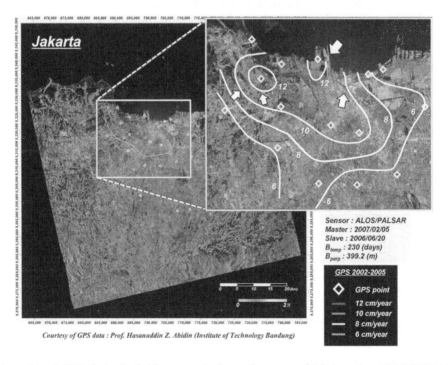

Figure 11.   InSAR-derived subsidence rate in the northern part of Jakarta using ALOS PALSAR data. The arrows are pointing toward the relatively large subsidence areas.

Please note in Fig. 10, that of the area showing the largest land subsidence is located in Pantai Mutiara housing complex. The GPS station MUTI is located in this area and, as shown in Fig. 7, this station also experienced large subsidence, i.e. about 80–90 cm during the period of December 1997 and September 2007.

6   LAND SUBSIDENCE AND GROUNDWATER EXTRACTION

Land subsidence in Jakarta is thought to be caused by four factors, namely: groundwater extraction, loading of buildings and other constructions, natural consolidation of alluvial soil, and tectonic movement. Up to now, there has been no information about the relative contribution of each factor to localized subsidence or their spatial distribution. In the case of Jakarta, tectonic movement is thought to be the least dominant factor for progressive subsidence, while groundwater extraction is considered to be one of dominant factors.

In the context of groundwater extraction, if the spatial distribution of land-surface subsidence in the period between 1982 and 1991 (Fig. 4) is compared with changes in the elevation of the groundwater piezometric surface (Fig. 12), it can be seen that a correlation exists. From this comparison it is suggested that the cones of depression within the piezometric surface inside the middle and lower aquifers more or less coincide with the cones of largest land subsidence measured by the levelling. In addition, each of the areas with the largest amount of land subsidence are situated in the areas consisting of sand bar and beach-river deposits; sediments that have high compressibility (Murdohardono and Sudarsono, 1998). Currently, these areas are industrial areas with relatively high-density settlement, both of which consume a lot of groundwater. This intensive groundwater abstraction appears to have deepened the piezometric water level inside the middle and lower aquifers and in turn caused land-surface subsidence above it.

The subsidence rate is closely related to the rate of piezometric water level (head) deepening in the middle and lower aquifers. According to Hadipurwo (1999), the maximum depth of the piezometric head inside the middle and lower aquifers of Jakarta tends to deepen with time, as shown in Fig. 13. In the case of Jakarta, the increases in both population and industry, which require a lot of groundwater, likely explain the declining trend

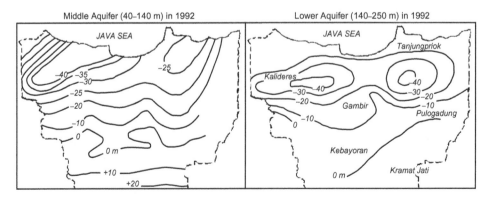

Figure 12.   Piezometric water level contours relative to MSL (in metres) inside Middle and Lower Aquifers of Jakarta in 1992; adapted from (Murdohardono and Tirtomihardjo, 1993).

of piezometric heads. This ever increasing demand on the groundwater resource appears to accelerate the deepening of piezometric head and, in a way, explains the higher maximum rate of subsidence in the period of 1991–1997 compared to those in 1982 to 1991 as observed by the levelling surveys (see Fig. 5). Moreover, Hadipurwo (1999) also observed that up to 1995 the depression cones of the piezometric heads tend to widen with time. This may explain the aforementioned increase in the number of subsidence cones in the period of 1991–1997 compared to those in 1982–1991 (see Fig. 4).

The groundwater level inside the Middle and Lower aquifers at several locations in Jakarta continue to decline. Fig. 14 shows that the groundwater levels are decreasing with rates of about 0.2 to 2 m/year over the period of 2002 to 2007. In comparison with GPS-derived subsidences shown in Figs. 7, 8 and 9, it can be seen that the large subsidences are usually associated with the relatively high rates of groundwater level change rates.

It should be realized however that in the shorter time scale, the groundwater level changes inside the Jakarta aquifers are quite dynamic, as shown by an example given in Table 1. These groundwater levels can go up and down up from several decimeters to a few meters in a year. The effect of this short-term variation in groundwater level inside the aquifers on the long-term subsidence phenomena in the Jakarta area and its spatial

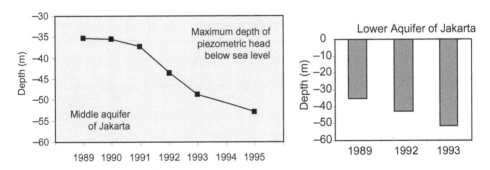

Figure 13.   The deepening of the piezometric head inside the middle and lower aquifers of Jakarta; drawn from the data given in Hadipurwo (1999).

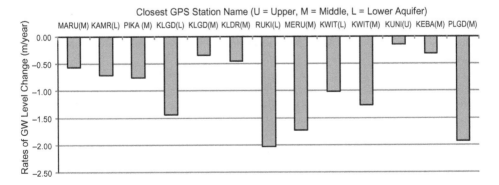

Figure 14.   Groundwater level change rates at several monitoring wells around certain GPS stations in Jakarta during the period of 2002 and 2007.

Table 1.   Example of variation in depth to groundwater for selected monitoring stations in Jakarta from 2002 to 2007.

Location	Closest GPS station	Aquifer	Groundwater level depths (m)					
			2002	2003	2004	2005	2006	2007
Cilincing (KBN)	MARU	Middle	−4.16	−3.42	−3.26	−3.94	–	−7.00
Kamal Muara	KAMR	Lower	−23.50	−23.36	−23.72	−25.62	–	–
Kapuk	PIKA, CEBA	Middle	−49.66	−49.82	−49.67	−50.73	–	−53.42
Sunter-1	KLGD	Lower	−11.11	−11.95	−12.57	−12.93	−11.75	−18.30
Sunter-3	KLGD	Middle	−21.38	−20.80	−20.82	−21.41	−21.08	−23.10
Tegal Alur	KLDR	Middle	−41.38	−39.47	−38.11	−41.98	−38.44	−43.65
Tongkol-5	RUKI	Upper	−4.71	−3.30	−3.29	−10.20	−4.19	–
Tongkol-7	RUKI	Lower	−25.13	−23.67	−22.47	−17.56	−23.54	−35.26
Tongkol-8	RUKI	Lower	−24.50	−23.67	−23.35	−27.67	−24.24	−28.85
Tongkol-9	RUKI	Middle	−8.73	−5.73	−6.21	−5.14	−7.12	−6.11
Joglo	MERU	Middle	−18.28	−20.15	−20.14	−24.48	−20.21	−26.92
Parkir Gd. Jaya	KWIT	Lower	−20.20	−20.60	−20.88	−21.86	–	−25.27
Atrium, Senen	KWIT	Middle	−12.26	−21.36	−21.66	−12.66	–	−18.56
Bapedalda Kngn	KUNI	Upper	−5.57	−5.42	−5.77	−4.56	−4.94	−6.28
Dharmawangsa	KEBA	92–125	−31.95	−33.29	−33.75	−32.26	–	−33.48
PT. Yamaha M-2	PLGD	Middle	−16.29	−12.51	−12.50	−13.54	–	−25.91
Pulogadung	PLGD	Middle	−28.13	−28.30	−29.46	−28.40	−28.47	−28.66

variations are not yet fully understood. More research is needed to study and clarify this matter.

7   LAND SUBSIDENCE AND SEA LEVEL RISE IN NORTHERN COAST OF JAKARTA

The levelling, GPS and InSAR derived results show that the coastal areas of Jakarta are affected by subsidence phenomena with rates of about 1 to 15 cm/year. During high tides, tidal flooding is already affecting some of these coastal areas. The extent and magnitude of subsidence related flooding will worsen with the likely continuation of sea level rise along the coastal area of Jakarta, which is bordered by the Java sea.

Fig. 15 shows the tide gauge data from Tanjung Priok station located close to the Jakarta harbour. The sea level rise trend is apparent in this tide gauge record from 1984 to 2004, with the rate of about 9 mm/year. This trend is also shown by the satellite altimetry results (see Fig. 16), with the slightly higher rate of about 15 mm/year. Based on these two data sets, it can be hypothesized that there is a sea level rise trend of about 1 cm/year in the coastal area of Jakarta. This sea level rise rate is much less than the subsidence rates of the coastal land area of Jakarta.

The combined effects of land subsidence and sea level rise in the coastal areas of Jakarta should be considered in vulnerability assessments of the areas to the tidal flooding phenomena. Table 2 shows two possible scenarios of future tidal flooding conditions, the

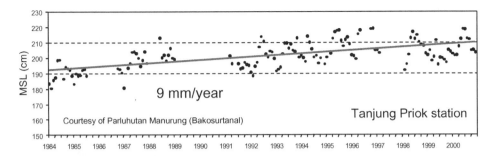

Figure 15.   Trend of sea level rise in the coastal area of Jakarta, as derived from tide gauge data; courtesy of Dr. Parluhutan Manurung, Bakosurtanal, Indonesia. The coefficient of determination (R^2) is about 0.85.

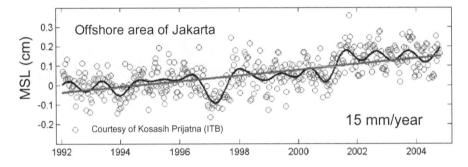

Figure 16.   Trend of sea level rise in the offshore area of Jakarta, as derived from satellite altimetry data (TOPEX/Poseidon and JAS-1), courtesy of Ir. Kosasih Prijatna MSc, KK Geodesy ITB, Indonesia. The coefficient of determination (R^2) is about 0.86.

first being a conservative estimate (most probable case) and the second being a pessimistic (worst case) scenario.

In the conservative scenario, a subsidence rate of about 2.5 cm/year and a minimum global sea level rise rate of about 2–3 mm/year (Gornitz, 1995; IPCC, 2007) are used. In the pessimistic scenario, a subsidence rate of about 10 cm/year and a local sea level rise rate of about 1 cm/year (as information from the tide gauge and satellite altimetry data) are used.

In the conservative scenario, the possible rise in sea level relative to the coastal areas of Jakarta could be up to 0.3 m in 2020 and 1.1 m in 2050, compared to its reference condition in the beginning of 2008. In the pessimistic scenario, these values would be a 1.3 m rise in 2020 and a 4.6 m rise in 2050. Considering the relatively flat nature (i.e. 0–2 m above MSL) of most coastal areas in Jakarta, this combined effect of land subsidence and sea level rise will certainly have disastrous consequences for habitation, industry, and fresh groundwater supplies from the coastal aquifers. Fig. 17 shows the possible inundated areas estimated using the scenarios given in Table 2.

It should be noted however that the land subsidence rate is not uniform over the entire coastal area of Jakarta, as shown in the previous Figs. 4, 7, 8, 9, 10 and 11. These Figures show some coastal areas are more susceptible to tidal flooding than the others. If the spatially different rates of subsidences as derived by InSAR (see Fig. 11) are integrated with the scenarios of sea level rise rates given in Table 2, then the possible inundation maps as given in Fig. 18 are obtained.

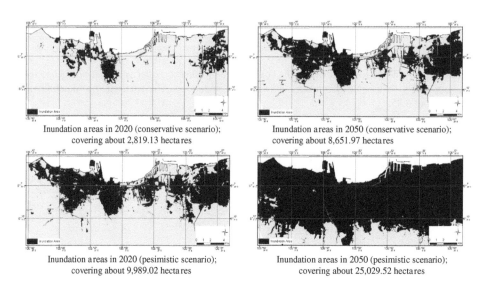

Inundation areas in 2020 (conservative scenario); covering about 2,819.13 hectares

Inundation areas in 2050 (conservative scenario); covering about 8,651.97 hectares

Inundation areas in 2020 (pesimistic scenario); covering about 9,989.02 hectares

Inundation areas in 2050 (pesimistic scenario); covering about 25,029.52 hectares

Figure 17.   Possible inundation areas (in black) in the coastal areas of Jakarta; with the assumption of homogeneous subsidence rates along the coast.

Table 2.   Possible combined effect of land subsidence and sea level rise in the coastal area of Jakarta.

Conservative Scenario	
Land subsidence rate	2.5 cm/year
Sea level (MSL) rise rate	0.2 cm/year
Separation rate between MSL and land surface	2.7 cm/year
Possible increase of sea level inundation in the coastal areas of Jakarta in 2020 (since beginning of 2008)	0.3 m
Possible increase of sea level inundation in the coastal areas of Jakarta in 2050 (since beginning of 2008)	1.1 m
Pesimistic Scenario	
Land subsidence rate	10 cm/year
Sea level (MSL) rise rate	1 cm/year
Separation rate between MSL and land surface	11 cm/year
Possible increase of sea level inundation in the coastal areas of Jakarta in 2020 (since beginning of 2008)	1.3 m
Possible increase of sea level inundation in the coastal areas of Jakarta in 2050 (since beginning of 2008)	4.6 m

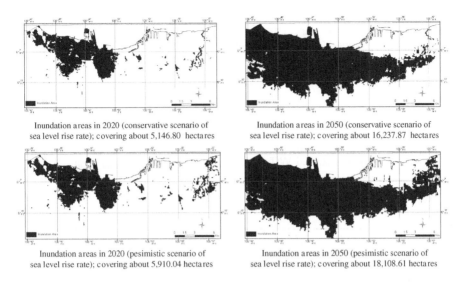

Inundation areas in 2020 (conservative scenario of
sea level rise rate); covering about 5,146.80 hectares

Inundation areas in 2050 (conservative scenario of
sea level rise rate); covering about 16,237.87 hectares

Inundation areas in 2020 (pesimistic scenario of
sea level rise rate); covering about 5,910.04 hectares

Inundation areas in 2050 (pesimistic scenario of
sea level rise rate); covering about 18,108.61 hectares

Figure 18.   Possible inundation areas (in black) in the coastal areas of Jakarta; with the assumption of spatially different subsidence rates along the coast.

8   CLOSING REMARKS

The results obtained from levelling surveys, GPS surveys and the InSAR technique over the period between 1982 and 2007 show that land subsidence in Jakarta has spatial and temporal variations. In general, the observed subsidence rates are about 1 to 15 cm/year, and can be up to 20–25 cm/year at certain locations and for certain time periods. There is a strong indication that land subsidence in the Jakarta area is related to the high volume of groundwater extraction from the middle and lower aquifers, with secondary contributions by building/construction loading and natural consolidation of sedimentary layers. The large volume of groundwater extraction causes a rapid decrease in groundwater levels inside the aquifers reducing the hydrostatic pressures on aquifer material, and in turn, causing the land surface above it to subside. However, the relation between land subsidence and localized groundwater level decrease will not always be a direct and simple relation, as indicated by example given in Figs. 19 and 20. It may be due to variations in the amount of groundwater being pumped over time. Therefore, if the amount of groundwater pumping is available, then it will be interesting to compare the correlation between the variations in pumping rates with the change in land subsidence.

Based on the data collected by the Provincial Mining Agency of Jakarta, it can be inferred that the groundwater level changes inside the aquifers of the Jakarta basin have rates of about a few dm/year up to a few m/year. Although most of the documented changes show a decline in the piezometric surface, some locations for specific aquifers and certain time periods also show groundwater level increases. In other words, short term variations in groundwater level also exist inside the aquifers of Jakarta basin. In studying the land subsidence phenomena in the Jakarta basin it would be interesting to investigate the effect of this short-term variation in groundwater pumping volumes and groundwater level on the long-term subsidence phenomena and its spatial variations.

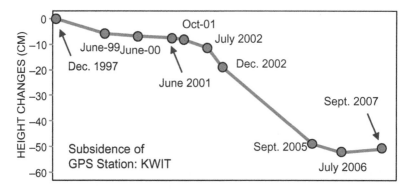

Figure 19.   GPS-derived ellipsoidal height changes of GPS station: KWIT.

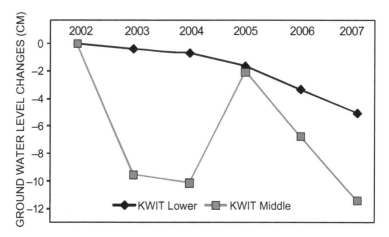

Figure 20.   Groundwater level changes observed at the monitoring wells around the KWIT GPS stations.

More data and further investigations are required to understand the intricacies of the relationship between land subsidence and groundwater extraction in the Jakarta basin. Additional causes of subsidence, e.g. load of buildings and construction, natural consolidation of alluvial soils, and tectonic movements, should also investigated and taken into account.

Finally it should be noted that in the coastal areas of Jakarta, the combined effects of land subsidence and sea level rise will introduce other collateral hazards, namely the tidal flooding phenomena. Several areas along the coast of Jakarta already have experienced tidal flooding during high tide periods. The adaptation measures to reduce the impacts of this phenomenon therefore should be developed as soon as possible.

ACKNOWLEDGEMENTS

The GPS surveys has been conducted using research grants from the Ministry of Science and Technology of Indonesia, Ministry of National Education of Indonesia, ITB Research

Grant 2005, and Kyoto University, Japan. The GPS surveys were conducted mainly by the staffs of Geodesy Research Division of ITB, and students of the Department of Geodesy and Geomatics Engineering of ITB. Ir. Bungaran Purba MM from the Provincial Mining Agency of DKI Jakarta is thanked for the information on groundwater level changes inside the aquifers of Jakarta basin.

REFERENCES

Abidin, H.Z. (2005). Suitability of Levelling, GPS and INSAR for Monitoring Land Subsidence in Urban Areas of Indonesia. GIM International. The Global Magazine for Geomatics, GITC Publication, Vol. 19, No. 7, July, pp. 12–15.

Abidin, H.Z. (2007). GPS Positioning and Apllications. In Indonesian languange. P.T. Pradnya Paramita, Jakarta. Edisi ke 3. ISBN 978-979-408-377-2, 398 p.

Abidin, H.Z., Djaja, R., Darmawan, D., Hadi, S., Akbar, A., Rajiyowiryono, H., Sudibyo, Y., Meilano, I., Kusuma, M.A., Kahar, J., Subarya, C. (2001). Land Subsidence of Jakarta (Indonesia) and its Geodetic-Based Monitoring System. Natural Hazards. Journal of the International Society of the Preventation and Mitigation of Natural Hazards, Vol. 23, No. 2/3, March, pp. 365–387.

Abidin, H.Z., Jones, A., Kahar, J. (2002). GPS Surveying. In Indonesian languange. P.T. Pradnya Paramita, Jakarta. ISBN 979-408-380-1. Edisi ke 2, 280 p.

Abidin, H.Z., Djaja, R., Andreas, H., Gamal, M., Indonesia K. Hirose, Maruyama, Y. (2004). Capabilities and Constraints of Geodetic Techniques for Monitoring Land Subsidence in the Urban Areas of Indonesia. Geomatics Research Australia. No. 81, December, pp. 45–58.

Abidin, H.Z., Andreas, H., Djaja, R., Darmawa, D., Gamal, M. (2008). Land subsidence characteristics of Jakarta between 1997 and 2005, as estimated using GPS surveys, GPS Solutions, Springer Berlin / Heidelberg, Vol. 12, No. 1: pp. 23–32.

Beutler, G., Bock, H., Brockmann, E., Dach, R., Fridez, P., Gurtner, W., Hugentobler, U., Ineichen, D., Johnson, J., Meindl, M., Mervant, L., Rothacher. M., Schaer. S., Springer. T., Weber. R. (2001), Bernese GPS software Version 4.2, In: Hugentobler, U., Schaer, S., Fridez, P. (eds) Astronomical Institute, University of Berne, 515 p.

BPS Jakarta (2007). Jakarta Dalam Angka 2007, Katalog BPS: 1403.31, Badan Pusat Statistik Propinsi DKI Jakarta, 520 p.

Deguchi, T. (2005). Automatic InSAR processing and introduction of its application studies. Proceedings the 26th Asian Conference on Remote Sensing, Hanoi, Vietnam.

Deguchi, T., Kato, M., Akcin, H., Kutoglu, H.S. (2006). Automatic processing of Interferometric SAR and accuracy of surface deformation measurement. SPIE Europe Remote Sensing, Stockholm, Sweden.

Gornitz, V. (1995). Sea-level rise: A review of recent past and near-future trends. Earth Surf. Proc. Landforms 20: 7–20.

Hadipurwo, S. (1999). Groundwater. In Coastplan Jakarta Bay Project, Coastal Environmental Geology of the Jakarta Reclamation Project and Adjacent Areas, CCOP Coastplan Case Study Report No. 2, Jakarta/Bangkok, pp. 39–49.

Harsolumakso, A.H. (2001). Struktur Geologi dan Daerah Genangan, Buletin Geologi, Vol. 33, No. 1, pp. 29–45.

Hutasoit, L.M. (2001). Kemungkinan Hubungan antara Kompaksi Alamiah Dengan Daerah Genangan Air di DKI Jakarta, Buletin Geologi, Vol. 33, No. 1, pp. 21–28.

IPCC (2007). Climate Change 2007: The Physical Science Basis, Summary for Policymakers, Report of Intergovernmental Panel On Climate Change, Can be accessed at: http://www.ipcc.ch/

Leick, A. (2003). GPS Satellite Surveying. John Wiley & Sons, Third edition, New York, ISBN 0471059307, 435 p.

Lubis, R.F., Sakura, Y., Delinom, R. (2008). Groundwater recharge and discharge processes in the Jakarta groundwater basin, Indonesia, Hydrogeology Journal, Springer, DOI 10.1007/s10040-008-0278-1.

Massonnet, D., Feigl, K.L. (1998). Radar Interferometry and its Application to Changes in the Earth's Surface. Reviews of Geophysics, Vol. 36, No. 4, November, pp. 441–500.

Murdohardono, D., Tirtomihardjo, H. (1993). Penurunan tananh di Jakarta dan rencana pemantauannya. Proceedings of the 22nd Annual Convention of the Indonesian Association of Geologists, Bandung, 6–9 December, pp. 346–354.

Murdohardono, D., Sudarsono, U. (1998). Land subsidence monitoring system in Jakarta. Proceedings of Symposium on Japan-Indonesia IDNDR Project: Volcanology, Tectonics, Flood and Sediment Hazards, Bandung, 21–23 September, pp. 243–256.

Purnomo, H., Murdohardono, D., Pindratno, H. (1999). Land Subsidence Study in Jakarta. Proceedings of Indonesian Association of Geologists, Volume IV: Development in Engineering, Environment, and Numerical Geology, Jakarta, 30 Nov.–1 Dec., pp. 53–72.

Rajiyowiryono, H. (1999). Groundwater and Landsubsidence Monitoring along the North Coastal Plain of Java Island. CCOP Newsletter, Vol. 24, No. 3, July–September, pp. 19.

Rimbaman, Suparan, P. (1999). Geomorphology. In Coastplan Jakarta Bay Project, Coastal Environmental Geology of the Jakarta Reclamation Project and Adjacent Areas, CCOP Coastplan Case Study Report No. 2, Jakarta/Bangkok, pp. 21–25.

Rismianto, D., Mak, W. (1993). Environmental aspects of groundwater extraction in DKI Jakarta: Changing views. Proceedings of the 22nd Annual Convention of the Indonesian Association of Geologists, Bandung, 6–9 December, pp. 327–345.

Sampurno (2001). Geomorfologi dan Daerah Genangan DKI Jakarta, Buletin Geologi, Vol. 33, No. 1: pp. 1–12.

Schreier, G. (Ed.) (1993). SAR Geocoding: Data and Systems. Wichmaann Verlag, Karlsruhe, ISBN 3-87907-247-7, 435 p.

Soetrisno, S., Satrio, H., Haryadi, T. (1997). To Anticipate Impacts of Reclamation of Jakarta Bay, A Groundwater Conservation's Perspective. Paper presented at Workshop on Coastal and Nearshore Geological/Oceanographical Assessment of Jakarta Bay: A Basis for Coastal Zone Management and Development, Jakarta, 25–28 June.

Yong, R.N., Turcott, E., Maathuis, H. (1995). Groundwater extraction-induced land subsidence prediction: Bangkok and Jakarta case studies. Proceedings of the Fifth International Symposium on Land Subsidence, IAHS Publication no. 234, October, pp. 89–97.

CHAPTER 11

Alluvial fans-importance and relevance: A review of studies by Research group on Hydro-environments around alluvial fans in Japan

Sung Gi Hu
Raax. Co., Ltd. Sapporo, Japan

Toshiki Kobayashi
Fukken Gijutsu Consul. Sendai, Japan

Eiji Okuda
IDOWR Engineering Co., Ltd. Tokyo, Japan

Akira Oishi
Yec. Co., Ltd. Tokyo, Japan

Mamoru Saito
Nippon Koei Co., Ltd. Tokyo, Japan

Tomoaki Kayaki
K-HGS Co., Ltd. Hiroshima, Japan

Seisuke Miyazaki
Yec. Co., Ltd. Fukuoka, Japan

ABSTRACT: We, the Research group on Hydro-environments around alluvial Fans in Japan, aim to study the present aquatic environment in and around the fans and how much they will be affected and changed by future temperature rise. The alluvial fans studied are as follows. Both Toyohira-gawa and Isawa-gawa alluvial fans are located in a rich snowfall area in northern Japan. Kurobe-gawa fan is situated in a heavy snowfall area in the middle part of Japan and faces the Sea of Japan. Inland basin and multi-use type Echi-gawa fan is located in an area having poor snowfall and faces lake Biwa-ko. Shigenobu-gawa fan is situated in the non-snowfall region in Shikoku Island, facing the Dohgo Plain. The Chikugo-gawa fan in the northern non-snowfall Kyushu region has no obvious features of an alluvial fan. For these six alluvial fans in Japan, this paper is mainly focusing on their topography, hydro-geology, water budget of each main rivers and interactions between surface water and groundwater and their future prospects.

Keywords: RHF, Toyohira-gawa, Isawa-gawa, Kurobe-gawa, Echi-gawa, Shigenobu-gawa, Dohgo Plain, Chikugo-gawa

1   INTRODUCTION

According to the 4th IPCC report (2007), the mean temperature on the Earth will increase by 6.4°C by 2100. Recently, it is observed that the mean temperature is continuously rising in various places of the world with a mean annual temperature rise of 0.6°C globally, 1°C in Japan and 3°C in Tokyo during the last 100 years. With this temperature rise, the IPCC also forecast that various unusual climatic changes will occur globally, for example, the number of typhoons will decrease but, in lieu, more powerful typhoons will devastate the land. Moreover, there will be an imbalance in precipitation and sudden local downpours will occur more frequently. Yet, the exact relationship between the temperature rise and the abnormal weather is not known, but it is a fact that unusual climatic phenomena are frequently occurring in the world.

In the summer season of 2003, an abnormal low temperature was experienced in Japan and a record heat wave swept through Europe, as a result over 30 thousand people died as a result of the heat wave. In 2004, the abnormal high temperature continued in the world and damages by the local heavy rains were enormous in many places in the world. Because of these abnormal climate changes, insurance premiums doubled compared to 2003. On the other hand, the record numbers of typhoons (ten) struck Japan in 2004.

In 2005, three out of the ten strongest hurricanes in the history of past 100 years devastated the Caribbean Sea coast of the USA. Especially, the Mississippi lowland and delta were wiped out by hurricanes Katrina, Rita and Wilma. In 2006, 14 typhoons struck Japan and, in the same year, winter witnessed the cold wave and had a sudden change to a warm winter in 2007. In 2007, heavy rainfall was recorded in Africa. Furthermore, unusual high-temperatures are frequently occurring in Siberia and Europe. Moreover, Japan did not experience any typhoon in 2008.

In this backdrop of climatic conditions, the main questions to be addressed include: a) what measures are to be taken during the sudden downpours or during the typhoon disasters that will be alternating in the near future, and b) how to prepare for the sudden downpours and/or flooding in terms of precautionary measures? Especially in this perspective, care should be taken with regard to how to secure the need for drinking water.

Today, disasters from low or rare rainfall are increasing every year in the other parts of the world, as the disasters from heavy rainfall. Leaving aside the perennially drought-prone Africa and Middle-East Asian countries that are continuously affected by drought, agriculture in Australia has been severely affected by drought and the lake near Sydney had been dried up due to low precipitation that is continuing for 7 consecutive years. Hence, with increasing number of drought-related disasters in the world every year a great number of people cannot get safe drinking water and this number is continuously increasing. Being a basic need for life, access to potable water for all people should be the issue of priority.

According to WHO, 1.2 billion people in the world have no secure access to safe drinking water today and the World Bank also reports that the number of these people will grow to 2 billion by the year 2020. The United Nations forecast that the number of people in the world will increase to 8.3 billion in 2025 and two-thirds of these people will not have access to enough safe fresh water. It is important and urgent that safe drinking water be available to those people.

After calculation using the SRES A2 scenario (CO_2 concentration is fixed to 860 ppm in 2100), the IPCC estimates that the mean temperature on the globe during 30 year period from 2071 to 2100 will rise by 3°C and that precipitation will also rise by 3.9% compared to that during the period of 30 years from 1961 to 1990. Japan's Meteorological Agency also forecast that the number of heavy rain days (over 100 mm per day) in the summer season in Japan will greatly increase during the period 2040 to 2100. On the other hand, the snow fall in the region along the Sea of Japan will be severely decreased until 2100, especially in the Hokkaido area, to two-thirds of its present volume (INOUE et al., 1998). A number of climatic and atmospheric institutes in the world have predicted that the recent climatic fluctuations will continue into the future due to the increase of greenhouse gasses.

Assuming that these predictions are correct, human civilization will face more threatening situations for fresh water resources and it will be difficult for mankind to avoid these for both fresh water and groundwater.

Thus, the risk management for the fresh water problems needs to be addressed well in advance. For example, the measurements for intense deluges, flooding and typhoon disasters are to be carried out to set the maintenance of infrastructure including the reinforcement of embankments, construction of artificial ponds, increasing the permeability of the land to precipitation. Similarly, for the drought it is required to preserve the forest and expand water retention capacities. Moreover, for securing safe drinking water, preservation and maintenance of the source of water supply and groundwater and their appropriate management will be required.

2   IMPLICATION OF GLOBAL WARMING ON LOCAL WATER BUDGET

If we assume that the global temperature may rise in future as is predicted, then the immediate issues that should be addressed include the effects on local surface water and groundwater. The Meteorological Agency of Japan (2005) demonstrate that the number of days with over 100 mm/d and 200 mm/d of precipitation have increased by a factor of 1.2 and 1.5, respectively, between 1901–1930 and 1975–2004. Moreover, days with over 100 mm/d of precipitation will increased by a factor of 1.5–2 by 2100 compared to the present. Drought is also evident in recent years as well as increasing heavy rain phenomenon. Unusual poor rain-days have increased two-fold since 1950 and predictions suggest future increases as well (simulation under the A2 scenario). With these anomalous future climatic conditions, it is important to identify the available potential of the freshwater resources to ensure its availability for the human civilization. Water, preserved in different aquifers underground and flowing on the surface mainly through the river system should be explored in a well-concerted manner so that its total balance and budget in restricted geographical and geological areas can be deciphered.

To enable adequate management of available groundwater resources, it is necessary to understand the natural conditions of the water basin such as hydro-geologic structure, recharge mechanisms, behaviour of the groundwater and water character. This includes, for example, monitoring the groundwater and accurate forecast of its use. It is needed to estimate the maximum permissible volume of groundwater from the structure of the water basin.

Groundwater systems have been an important part of the hydrological cycle of the Earth. For sustainable use and development of groundwater, adequate and proper management of the fresh water including groundwater is a major need. Moreover, groundwater is especially sensitive to changes relate to the climatic conditions as it is intricately related to the atmospheric temperature, precipitations and anthropogenic activities among other parameters. This necessitates a detailed scientific understanding of the groundwater and fresh water system that is quite indispensable for its general and crisis management.

3   REASONS BEHIND CHOOSING ALLUVIAL FANS

An alluvial fan is a fan-shaped deposit formed where a fast flowing stream flattens, slows, and spreads typically at the exit of a mountain area onto a flatter plain (SAITO, 1988). In approaching the methodological challenges of defining the water balance of groundwater, there are several spatial options to work with. One is the large flat plain terrain where the initial problem is to decide over the groundwater divisions. Second is the case of restricted well-defined basin areas in between the mountains. Alluvial fans are moderately wide and well-defined areas from geological, geo-morphological and hydrological point of view. Hence, they are the most challenging in terms of our main aim to study the water balance problems.

That is to say alluvial fans:

1. Shows a moderate extent of land and are considered to be a closed system from the geological point of view,

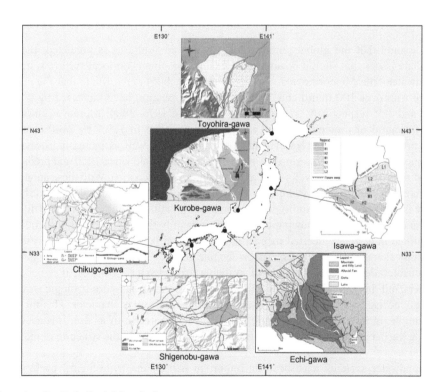

Figure 1.   Studied alluvial fans in Japan.

2. Have a simple interaction between the surface water and the groundwater, and it is easy to understand the relations of recharge and discharge in the fan,
3. Form an area where it is easy to establish the framework of hydrological boundaries for simulation and the demarcation of watersheds,
4. Are geologically, topographically, hydro-logically and hydro-geologically well-studied,
5. The problems of water in such fans are closely related to the activities of human civilization, culture and economical activities,
6. Groundwater flows relatively rapidly underground and is adequate to make simulations for the future climatic changes,
7. Are widely spread not only in Japan but also all over the world.

Thus, the alluvial fan shows peculiarity but universality in its character. Carrying out concrete investigation and simulation for the effects of global warming to clarify the mechanism of ground-water recharge, interaction of both surface water and groundwater, and the balance of groundwater will lead to understand the present state of water balance of the fan. In the backdrop of future conditions this is the main aim of our effort through RHF (**R**esearch group on **H**ydro-environment around alluvial **F**ans in Japan).

In this regard, we have chosen a series of studies that relate the fluctuation of water and its balance under the effects of climatic changes, especially the projected temperature rise in the near future due to global warming.

The RHF is studying various kinds of alluvial fans. The alluvial fans studied in Japan by RHF, with the aim to understand the mechanism of groundwater recharge, interactions of both surface water and groundwater, and the present state of the water balance of fans are as follows (Fig.1). Both Toyohira-gawa fan (Sapporo, Hokkaido) and Isawa-gawa fan (Iwate Prefecture) are situated in a rich snowfall area in northern Japan, Kurobe-gawa fan (Toyama Prefecture) is situated in a heavy snowfall area in the middle part of Japan and faces the Sea of Japan. Echi-gawa fan (Shiga Pref.) is an inland basin type and is located in a poor snowfall area. Both Shigenobu-gawa fan (Ehime Pref., Shikoku Island) and Chikugo-gawa fan (Fukuoka Pref., Kyushu) are in a non-snowfall region (Japanese "-gawa" means river).

4   CHARACTERISTICS OF STUDIED ALLUVIAL FANS

Generally, the riverbed slope from the mountain area to the sedimentary basin is certainly steep and the force of erosion and transportation of the river is severe in Japan. So, the gravel to cobbles are dominantly distributed in the proximal fan, while fine-grained sand to muddy materials occupy the distal fan.

The surface water usually permeates into the ground at the proximal to mid-fan part (for all the studied fans). Some of the rivers usually drastically loose their water during flow down the riverbed (e.g. Echi-gawa fan, Shigenobu-gawa fan and Chikugo-gawa fan area), which emerges at the break of slope as non-pressured groundwater (e.g. Echi-gawa fan, Shigenobu-gawa, Chikugo-gawa fan) and water also gushes out at the end of the fan as a non-pressured or pressured groundwater forming a discharge zone of the alluvial cone (e.g. Toyohira-gawa and Echi-gawa fans), and finally, a remnant groundwater directly flows out to the sea or lake bed (in case of Kurobe-gawa and Echi-gawa fans). Both Toyohira-gawa and Echi-gawa fans consist of several aquifers and have multiple spring zones due to the complicated formation processes in such alluvial fans.

Table 1. Compare the six alluvial fans in Japan.

Name of the Fan	River name & its scale*1	Location of the Fan	Scale*2	Geology & Structure	Hydro-geology (Thickness,max.m)	Permeability (m/s)	Meteorology*3	Volumes*4 (million m³/y)	G.W. Storage (million m³/y)	Utilization	Type of the Fan	Characteristics & Remarks
Toyohira-gawa (Hokkaido)	Toyohira-gawa L=72.5km R=894.7km²	N43° 04'07" E141° 21'03"	R=7km L=7km H=45m I=7/1,000 E=35km²	Holocene Pleistocene Pliocene	Aquifer I (20) Aquifer II (30) Aquifer III (70) Aquifer IV (40) Marine deposits	2×10⁻⁵, 6×10⁻⁴ 1.2×10⁻⁵ 9×10⁻⁶ 4×10⁻⁷	T = 8°C P = 1,090mm ET = 630mm S = 630cm (Total depth) S = 101cm (Max. depth)	D = 882 I = 642 If = 204	300	Urban	Rich snowfall region Composit type(Pleist.+Holoc.) Semi-closed type Urban use	Recharge from Riv. → G.W. Surface → G.W. G.W. → Riv. Menu (= spring) ; vanished 1.9 million population
Isawa-gawa (Iwate pref.)	Isawa-gawa L=44.3km R=190km²	N39° 07'09" E141° 03'08"	R=18km L=23km H=80~100m I=5/1000 E=110km²	Holocene Tarrace Pliocene Miocene Paleozoic	Holocene(10) Terrace Pliocene(40~120) Miocene	2.5×10⁻⁴ 1~10×10⁻⁵ 1×10⁻⁵	T = 8°C P = 1300mm ET = 682mm S = 112cm	D = 157 I = 0 If = 152	620	Agriculture	Rich snowfall region 7 inclined terrace surface Semi-closed Type Agricultural use	Recharge from precipitation, snow melts and artificially No direct recharge into the ground from the River Recharge from surface → G.W.
Kurobe-gawa (Toyama pref.)	Kurobe-gawa L=85km R=682km² I=1/40	N36° 51'40" E137° 33'10"	R=13.6km L=13.6km H=130m I=1/100 E=120km²	Holocene Pleistocene Tertiary	Holocene(20-80) Pleistocene(60~160) (Bae,1980)	2×10⁻⁶ 2×10⁻⁷	T=14°C (1971~2000) P=2,167mm (end) (1981-2005) P=3,605mm (top) (1971-2000) ET=780mm (1971~2000) S=295cm (Total depth) S=54cm (Max. depth)	D = 2,551 I = 890 If = 440	2,000~3,000	Agriculture Industry Snow melting	Faced to the Sea Multi-used Fan Open type Heavy snowfall region Marine dep.+Fluvial dep.	Segire*5 Spring out from the seabed or end of the fan Recharge from the Riv. → G.W. Surface → G.W.
Echi-gawa (Shiga pref.)	Echi-gawa L=53km R=204km²	N35° 06'48" E136° 12'48"	R=15km L=15km H=100m I=7/1000 E=98km²	2 Holo.Ter. 3 Pleist.Ter. Ko.Biwako G. (Plio.-Pleistocene)	Shallow Aquif.(20) Interm. Aquif.(40) Deep Aquif.(40>)	1×10⁻⁴~10⁻⁵ 1×10⁻⁵~10⁻⁶ 1×10⁻⁴~10⁻⁶	1979 2006 T=13°C T=15°C P=1500mm P=1300mm ET=770mm ET=800mm	D = 113 I = 233 If = 358	1,360	Agriculture Tap Water	Rare~non snowfall region Pleistocene Aquifer River flows to the Lake Biwa-ko Semi-closed type	Segire dominant Springs in the lake bottom (~20m) Discharge from shallow lake bottom to the lake Biwa-ko
Dougo Plain (Ehime pref.)	Shigenobu-gawa L=36km R=443km² L=16km (Plain) I=9~13/1,000	N33° 50'36" E132° 46'36"	R=4.3km L=4km (Shigenobu-gawa Fan) I=13.1/1000 E=18 km²	Holocene~ Pleistocene Cretaceous	Shallow Aquifer Deep Aquifer	2×10⁻⁵~10⁻³ 1×10⁻⁵~10⁻⁴	T = 16°C P = 1,300mm ET = 870mm (1974-2003)	D = 221 I = 78 If = 143-85 (143=recharge) (85=P-ET in the plain)	110~1,540	Agriculture Tap Water Grove	Non snowfall region Two-storied Basins Closed & stagnant type Agriculture and Grove	Segire dominant Three flow ways of G.W. Spring= Izumi (agricultural use) rich 600 thousand population
Chikugo-gawa (Fukuoka pref.)	Koishiwara-gawa Sata-gawa	N33° 25'42" E132° 38'54" N33° 24'22" E130° 28'00"	R=5km L=5km I=6/1,000 E=45km² R=5km L=10km I=6/1,000 E=125km²	Pleist.~Holo. Mioc.~Plioc. Metamorphics. & Granite	Aquif. I (~20) Aquif. II (20)	3~6×10⁻⁴ 2×10⁻⁵	T = 16°C P = 1,870mm (1976-2006) ET = 870mm	D = 148 I = 222 If = 74	1,000	Agriculture Urban	Non snow fall region Multi(two old & one new typed) fan Semi-closed & vanishing type Agriculture (main)	Segire Recharge from precipitation, river and groundwater

*1 L:Length, R:Reservoir, I:Inclination
*2 R:Radius, L:Length, H:Relative Heights, I:Inclination, E:Extent
*3 T:Temperature, P:Precipitation, ET:Evapotranspilation, S:Snow
*4 D:Discharge, I:Inflow vol, If:Infiltration vol.
*5 "Segire" means river waters are completely permeated into the ground and as a result, surface water could not seen on the riverbed for a long time intervals. (= interrupted stream)

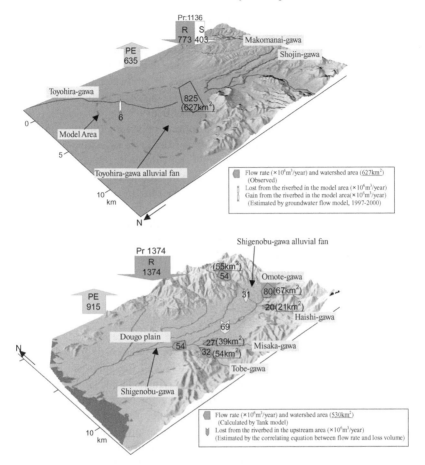

Figure 2. Comparison of the hydrological environments of Toyohira-gawa alluvial fan and Dohgo Plain. Pr: Precipitation (mm/year), R: Rainfall (mm/year), S: Water equivalent of snow fall (mm/year), PE: Potential evapo-transpiration calculated by the Thornthwaite method (mm/year) (1997–2000 Average).

A large amount of surface water on the alluvial fan flows into the underlying aquifer so that the rivers always have a low water volume except for the flooding time (e.g. Echi-gawa fan and Shigenobu-gawa fan in Dohgo Plain) and the ephemeral riverbeds are usually formed on the fan (e.g. Dohgo Plain). Under these natural conditions, it is relatively easy to estimate the water balance.

Although alluvial fans have a complicated character relating to their groundwater and surface water due to a great variety of sedimentary formation processes, alluvial fan are of reasonably extent yet intermediate in size, and can be considered as a closed system for both groundwater and surface water. Due to this reason, it is easier to understand the groundwater basins of the alluvial fans than those of other basins such as in the Quaternary plains.

At present, knowledge of the real volume of groundwater in each of the alluvial fans under investigation is meagre. However, the groundwater volumes in each fan have been estimated, e.g. as 300 million m^3 in Toyohira-gawa fan, 2 to 3 billion m^3 in Kurobe-gawa fan, 1.3 billion m^3 in Echi-gawa fan and 110 million to 1.5 billion m^3 in Dohgo Plain, respectively (Table 1). Moreover, because a large amount of the water in the alluvial fans usually exists at shallow depth and as a non-pressured groundwater (unconfined aquifer), it is easy to use for industry, agriculture and sewage disposal among others (as for all the alluvial fans except for Toyohira-gawa fan). Thus groundwater in the fan, as shown in Table 1, has close relationships with anthropogenic activity, particularly where ground-water can be easily accessed from multiple aquifer beds.

Despite the vast volume of groundwater resources existing at the present time, but it is necessary to know the correct volume in the fans from these investigations and also it will be useful to forecast the fluctuation of groundwater volume due to changes in the recharge volumes in the future. Ground- water resources will become the most important one in the near future due to changing precipitation caused by the global warming. So, the result is of great significance for future generations.

From this perspective, the representative case studies in two of the studied fans i.e., Toyohira-gawa fan in a rich snowfall region and Shigenobu-gawa fan in a non-snow fall region are shown schematically in Figure 2. Detailed results for each alluvial fan and other fans are available in this volume.

5   EPILOGUE

While planning and deciding over the methodology of evaluation of groundwater in alluvial fans with its present problems, we would like to recall the aphorism in 1980's society of young people in the west coast of America, saying "Think Globally, Act Locally". Firstly, on a global scale, the most important thing for us is to put a brake on further deterioration of the climate conditions. At a local scale, it is important to develop unified administration/management for both the surface water and groundwater. For this, it is necessary to understand precisely the present local conditions of the fresh water. From a technical point of view, we have to get hold of the division of the volumes of surface flows and recharge, get a clear picture of the structure of aquifer beds, know the volumes of natural spring water, base-flow, the volume of abstraction and the real state of utilization of groundwater in much detail.

Today in Japan, fresh water administration is looked after by the central ministry or local authority, but the groundwater does not belong to their purview. As groundwater problems are as equally important as those of surface water, integrated administration of both sur-face water and groundwater is urgently needed. However, politically it is most important to exclude this administrative organ from their thinking of narrow territorial awareness. And socially, it will be important to reform consciousness (by means of education, announce-ment and publicity) among the citizens that the fresh water resource is the *common* and limited treasure for life.

ACKNOWLEDGEMENT

The committee of RHF had met 6 times since 2004 at various places where the targeted fans are located. The RHF is appropriately administrated by the guidance of committee. We, at the secretariat of RHF, wish to express hearty thanks for the committee for their continuous encouragement and valuable advice throughout this work. We also express our hearty thanks to the staff of the River Development and Construction Departments of Hokkaido Development Bureau and Regional bureaus of Tohoku, Hokuriku, Shikoku and Kyushu Ministries of Land, Infrastructure and Transport for their helpful support through our study and works.

Thanks are also due to the members of RHF for their cooperation through this study.

REFERENCES

Hu, S.G. (2008) Why have we picked up alluvial fan-Importance and Relevance. RHF (ed.) Hydro-environments of Alluvial Fans in Japan Monograph, *36th IAH Cong. 2008 Toyama*, 5–10.
Inoue, S., Akiyama, K. (1998) Change Prediction for Snowfall at the time of Global Environmental Changing (in Japanese).
IPCC WG 14th Assessment Report (2007) Climate Change 2007: The Physical Science Basis. 976 p.
Meteorological Agency (2005) Report on unusual weather (in Japanese). http://www.data.kishou.go.jp/climate/cpdinfo/climate_change/
Saito, K. (1998) Fans in Japan (in Japanese). Kokon Shoin, 280 p.

CHAPTER 12

Study on the relation between groundwater and surface water in Toyohira-gawa alluvial fan, Hokkaido, Japan

Sung Gi Hu
Raax Co., Ltd. Higashi-ku, Japan

Shigechika Miyajima
Hokkaido Regional Development Bureau, MLIT, Chuou-ku, Japan

Daisuke Nagaoka, Ken Koizumi & Kazuyuki Mukai
Raax Co., Ltd. Higashi-ku, Japan

ABSTRACT: The Toyohira-gawa alluvial fan is an urbanized fan. A billion tonnes of river runoff flows every year on the surface of the fan, and is the source of tap water to the 1.9 million people in Sapporo region. Moreover, there are also 300 million tonnes of groundwater stored in the fan. The river Toyohira-gawa is recharged by both rain and snow, in approximately equal amounts. In addition the fan is recharged from the river water. Recharge from the riverbed to the underground is estimated as 321mm/y, while 277mm/y is from the land surface. Though the volume of discharge from the groundwater to the river cannot be estimated, it is apparent that there is an active interaction between surface water and groundwater. This is evident from several losing and gaining water segments on the riverbed.

Keywords: Toyohira-gawa alluvial fan, recharge, discharge, interactions, circulation, water balance

1   INTRODUCTION

Generally, the alluvial fan shows universality in its character and distribution in spite of some unique characteristics of each fan. The present study was carried out to understand the mechanism of the groundwater recharge system, interaction of both surface water and groundwater and the balance of groundwater that would lead to understanding the present state of the water balance of the Toyohira-gawa alluvial fan, one of the typical fans in Japan.

The river Toyohira-gawa (hereafter Toyohira-gawa, or the River) is one of the main first stage tributaries of the river Ishikari-gawa, the biggest river in Hokkaido, which flows from south to north through Sapporo City (N43°04′07″ E141°21′03″). The River is 72.5 km long and has a 894.7 km^2 catchment area. Toyohira-gawa alluvial fan is a complex fan (Daimaru, 1989) which is heavily urbanised. The fan was formed at the end of the Pleistocene to early Holocene period, with the riverbed inclined in a gentle slope at the downstream of the River. This fan is 7 km in radius, 7 km wide at the distal part of the fan, has a 45 m relative height

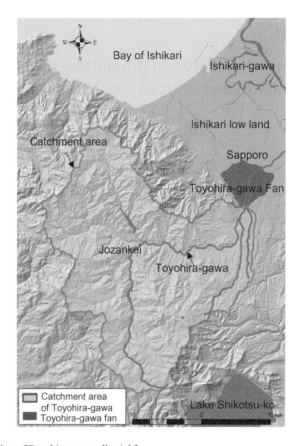

Figure 1.   Location of Toyohira-gawa alluvial fan.

differences between the distal and proximal parts of the fan and extends over an area of 35 km² (Fig. 1).

This paper primarily focuses on the hydro-geologic structure of the fan, deciphering the water circulation by analyzing the water quality, stable- and radio-isotopic characters, ^{14}C ages and finally, estimates the water budget of Toyohira-gawa alluvial fan and the River, taking into account the interactive relationships between groundwater and surface water in the fan.

2   HYDRO-GEOLOGIC STRUCTURE AND GROUNDWATER FLOW

2.1   *Geomorphology and geology*

Toyohira-gawa alluvial fan is a complex fan that extends onto both the Hiragishi surface, at an elevation of 90 m–20 m above sea level and formed during the upper Pleistocene, and the Sapporo surface which is 50 m–15 m in height and formed in the Holocene. The fan is surrounded by the Neogene Tertiary lava formations (200 m–500 m in height) in the west, with small-scale alluvial fans in between the lava and upper Pleistocene volcanic

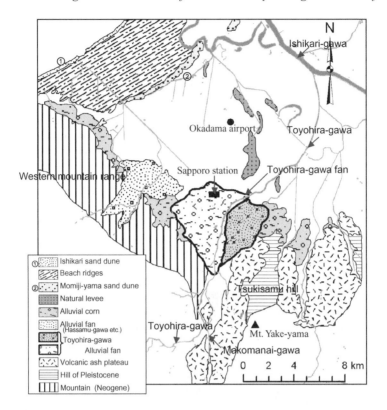

Figure 2.   Geological map around the fan.

pyroclastic terrain with 200 m–100 m in height in the east. Tertiary hyaloclastic formation and Quaternary volcanic pyroclastics are present in the south and Holocene marine to fluvial deposits (5 m–1 m in height) are present in the north (Fig. 2). The mean inclination of the Hiragishi surface and the Sapporo surface are estimated to be 9/1,000 and 7.5/1,000, respectively and the average inclination of the riverbed of Toyohira-gawa, which flows on the fan, is 7/1,000. The basement of the fan consists of Pleistocene marine deposits.

Six fluvial terraces, having relative heights of 40 m–5 m, can be seen on the left and right sides of the Toyohira-gawa from Jozankei to Sapporo. One of these fluvial terraces (namely t3) continues to the Hiragishi surface and the other (t5) extends until the Sapporo surface (Fig. 3). Judging from the relationship between the fan surface and the six fluvial terrace plains, it seems that the riverbed of Toyohira-gawa has shifted from east to west. Initially, the Toyohira-gawa was flowing more to the east than today's riverbed, eroding the Shikotsu volcanic ash deposits and forming the Hiragishi surface in the last glacial age (Hu, 2008; Nagaoka et al., 2008). Finally, the Toyohira-gawa resorted to its present riverbed in the post- glacial Holocene period (Fig. 4).

The Toyohira-gawa alluvial fan consists of both Pleistocene and Holocene deposits of sand, gravels and silts. The Pleistocene Hiragishi surface consists of three sedimentary strata. The lower strata of the Hiragishi surface deposits are thin alternations of clayey silt

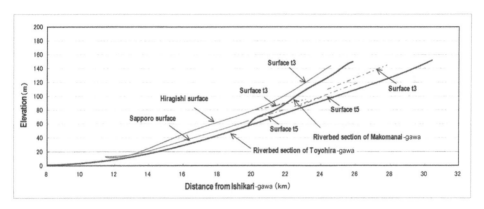

Figure 3.   Riverbed sections of Toyohira-gawa and Makomanai-gawa with the alluvial fan surface and the terraces.

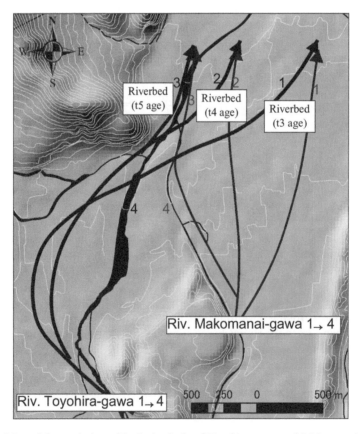

Figure 4.   Map of the evolution of both riverbeds of Toyohira-gawa and Makomanai-gawa. The rivers changed their course from 1 to 4 during 18,000 to10,000 years before present. In 18,000 years before present, the two rivers joined at place 1.

and sand of more than 20 m in thickness, while the middle strata is mainly composed of gravels and can be sub-divided into two beds on the basis of their permeability—a lower sand/gravel bed of 30 m–5 m in thickness and an upper sand/gravel bed of 30 m–15 m thickness with several sandwiched thin silt layers in the upper strata. All of the Hiragishi surface strata consist of pebble to cobble sized andesitic rocks within a coarse-grained sand matrix. At the distal part of the fan, Holocene sand and silt beds overlie it. The Holocene Sapporo surface of the fan also consists of sand/gravel beds with andesite cobbles with silty clay or with sporadic intercalations of irregular sized thin sand beds. The matrix of the sand/gravel bed is made up of coarse-grained sand.

2.2 Hydro-geologic structure

The marine sediments, which correspond to the Pleistocene Nopporo Formation, are considered as the hydro-geological basement strata of the fan. They are overlain in ascending order by the uppermost Pleistocene formations, i.e. no. IV aquifer consisting of sand and gravel bed (maximum 40 m in thickness) and no. III aquifer (maximum 70 m in thickness), as well as the Holocene no. II aquifer (maximum 30 m in thickness). All of these aquifers have intercalating thin low-permeability to impermeable strata (2 m–10 m in thickness) and have a gentle slope from the southern upper part to northern lower part of this fan. In addition, the Holocene no. I aquifer consists of sand, clayey beds and peaty beds of 10 m to 20 m in thickness on the northern part of the fan (Fig. 5). The total groundwater capacity for these hydro-geologic structures of the fan, derived from the effective porosity estimated from the drilling cores, is estimate to be 300 million m^3. The contour map of the base of the main aquifers (no. II and III) reveals that no. II aquifer is inclined to the northwest (left in Fig. 6). On the other hand, no. III aquifer shows bi-directional inclinations, i.e. one from the southwest mountain area to the central part of the city in the Sapporo surface, while in the Hiragishi surface, the inclinations are from the proximal fan to the northeast direction and to the Toyohira-gawa directions (right in Fig. 6).

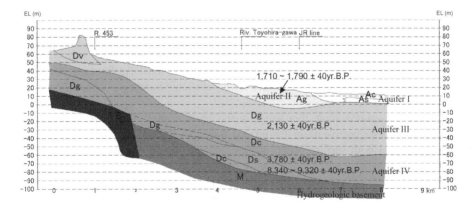

Figure 5.   Hydro-geological profile of Toyohira-gawa fan.
Abbreviations: A; Holocene, D; Pleistocene, B; Tertiary bedrock, M; Hydrological basement strata (mud stone), v; pyroclastics (volcanic ash etc.), s; sand, g; gravel, c; silt to clayey. [14]C ages of groundwater in each aquifer are also shown in the Fig. (see text).

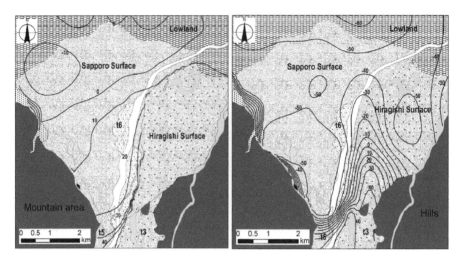

Figure 6.   Contours (m) of bottom surface for each main aquifers (left; Aquifer II, right; Aquifer III). t3, t5, Terrace surface.

2.3   *Permeability*

The permeability of no. IV aquifer, which has a thickness of 5~40 m, is 4×10^{-7} m/s; that in no. III aquifer (10~70 m thick) is 9×10^{-6} m/s; in no. II aquifer (10~30 m thick) is 1×10^{-4} m/s; and in no. I aquifer (10~20 m thick) is 2×10^{-5} m/s~6×10^{-4} m/s. There are no permeability tests on the clayey bed that intercalates in the aquifer, but the permeability is estimated as 1×10^{-8} m/s.

2.4   *Presumptive groundwater flow directions by water quality*

The results of the water quality analyses of the surface water and water from each aquifer are shown in Fig. 7 on ternary diagrams.

Aquifer II has a calcium bicarbonate-type water, including nitrate-nitrogen in places. Aquifer III also has a calcium bicarbonate type water, although it has an inferior ionic balance due to low concentrations of positive ions. Some of the samples indicate a lack of calcium ions and/or low concentrations of magnesium ions. As far as aquifer IV is concerned, the sodium bicarbonate waters have poor ionic balances, especially for both calcium and magnesium ions.

The ternary diagrams show that water in the aquifers plot around the centre of the diagram, which is suggestive of an unconfined water circulation field. This also indicates that the quality of both surface water and groundwater on the fan have no conspicuous characteristics.

Groundwater flow occurs in two directions i.e. (1) from aquifer no. II (mainly consisting of Holocene gravels) to no. III (Pleistocene gravels) and (2) from aquifer no. II (Holocene gravels) to no. IV (Pleistocene sand) via no. III (Pleistocene sand) from the proximal to the distal fan (Fig. 15). Two patterns of water quality changes are evident from Fig. 15. Generally, the water in the Toyohira-gawa alluvial fan is of the calcium bicarbonate type. In the former flow direction (case 1), water quality does not change to any other type and

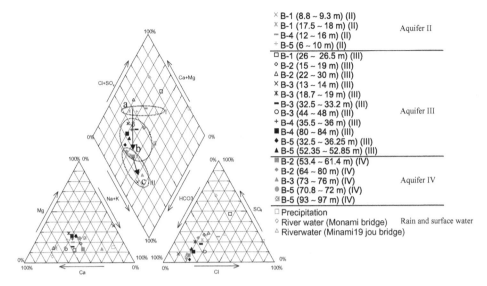

Figure 7. Ternary diagram showing rain, surface water and groundwater quality. Solid line with arrow in the key diagram represents the water quality changes from a to c.

so they plot in the same field in the ternary diagram. In the latter (case 2), water quality changes remarkably from a calcium-rich bicarbonate type to calcium-poor type and from calcium-poor to non-calcium bicarbonate type by ion-exchange during passage through the different geologic strata and aquifers.

As described above, water quality in case 1 plots in a narrow and well-defined area in the ternary diagrams, which is the same as the usual unconfined groundwater. Whereas in case 2, the starting field is the same as case 1, but during groundwater flow to the different aquifers, both sulphate and chloride anions and calcium and magnesium cations are decreased by 30% and 20%, respectively. With further passage through the aquifers, the cations decreased further by 20% and sodium, potassium and bicarbonate ions increased (Fig. 7). The change in water quality along the groundwater flowpath to the river could not examined yet. However, it is assumed that groundwater quality will transform to the calcium bicarbonate type in the discharge zone because the river water has a calcium bicarbonate type, while the discharged groundwater has calcium-poor to non-calcium bicarbonate types and its estimated volume is too small compared to the surface water.

2.5 Origin and ages of groundwater

Stable isotope, radioisotope and ^{14}C age analyses on groundwater were carried out to unravel the origin of the groundwater of the fan. According to the delta diagram (Fig. 8), the δ values of the groundwater in the fan are comparable to the δ values of the groundwater of Shikotsu volcanic ash plateau at southern part of the Toyohira-gawa alluvial fan. All the samples of groundwater from the fan plotted in a narrow field in the diagrams, i.e. δD between $8\delta^{18}O+13$ and $\sim 8\delta^{18}O+19$, except for one sample. Though samples of the fan are relatively lighter than those in the Shikotsu volcanic ash plateau, both the samples from

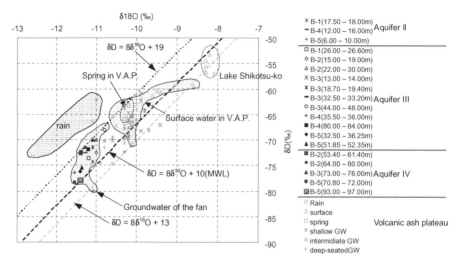

Figure 8.   Delta diagram showing the various types of water around the fan. Delta values of the fan samples are the lightest in the diagram. Values of the samples of lake Shikotsu-ko are probably condensed during stagnant time. Note that some delta values of deep groundwater in the volcanic ash plateau (V.A.P) are plotted in the same area as that of the fan.

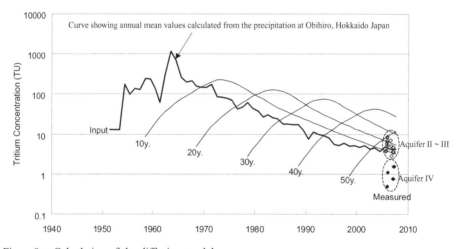

Figure 9.   Calculation of the diffusion model.

the fan and the volcanic ash plateau plot above the line of Meteoric Water Line (MWL) ($\delta D = 8\delta^{18}O + 10$) which indicated that originate from precipitation (Takahashi et al., 2003).

Tritium concentration values in aquifers II, III and IV shows that there is no difference in tritium concentrations for both waters in aquifers II and III with mean values of 4.5 T.U. and 5.6 T.U., respectively. On the other hand, tritium in aquifer IV is measured as 1.0 T.U. (mean value) which indicates very old water in this fan. Calculations of the change of tritium concentrations using a diffusion flow model shows a 30 year stagnation period of groundwater in aquifers II and III, and 50–55 years in aquifer IV (Fig. 9).

^{14}C ages of groundwater in the aquifers at various depths were analyzed (Fig. 5). Samples taken in aquifer II (at depths of -17.5 m and -26 m below ground level) are estimated as $1,710 \pm 40$ yr B.P. and $1,790 \pm 40$ yr B.P. Water in aquifer III (at -35.5 m depth) is of $2,130 \pm 40$ yr B.P. and in aquifer IV (taken at depths of -52.5 m and $-71\sim-93$ m) is of $3,780 \pm 40$ yr B.P. and $8,340\sim9,320 \pm 40$ yr B.P., respectively. The data shows that the ^{14}C ages of groundwater increase with depth, although there are no clear differences between aquifers II and III.

3   WATER CIRCULATION AND ITS BALANCE AROUND THE FAN

3.1   *Hydrological and meteorological analyses*

The mean annual temperature of Sapporo city is 7.8°C (mean monthly temperature from 1901–2000 for August and January are 21.5°C and -5.2°C, respectively), mean annual precipitation is 1,013 mm (1901–2000) and the average annual value of actual evapo-transpiration is 630 mm, calculated by the Thornthwaite method based on annual values from 1996–2000. The mean temperature during the past 100 years has increased by 2.35°C (January, 2.97°C; August, 1.52°C), although the values are higher than that of the mean temperature increase of Japan due to the heat-island phenomenon. However, mean annual precipitation gradually increased by 162 mm from 1900 to 1980. In 1984, though, it decreased to the level of 1900 but is again increasing in recent years (a 75 mm increase during last the 23 years compared to 1984, Fig. 10).

Incidentally, the maximum annual precipitation in Sapporo area during the past 100 years was recorded in 1981 as 1,672 mm/y which caused huge flooding in the same year.

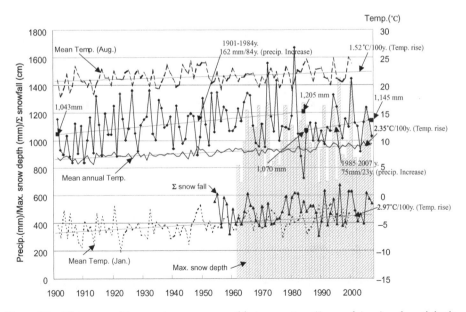

Figure 10.   Mean annual temperature, mean monthly temperature (Jan. and Aug.) and precipitation (rain, amounts of snowfall and maximum snowfall) in Sapporo during 1900 to 2010.

In contrast, three years later in 1984, the minimum annual precipitation of 725 mm/y was recorded.

3.2 *Discharge and water balance of the river*

Major increases in the flow of the river Toyohira-gawa occurs twice in a year, once during the snow melting season in spring and the other during the rainfall season in autumn, whereas shortage of water is experienced during the summer and winter. These two cycles represent one hydrologic year. In Toyohira-gawa, maximum runoff is recorded between April and May, and minimum is between August and September (Fig. 11).

River runoff in Toyohira-gawa is altered by gaining and losing water from/to the under-lying aquifer at several places throughout the year (Fig. 12). During the dry season in the river, water infiltrates into the ground at the uppermost part of the fan and discharges back into the river several hundred meters downstream. Waters are repeating this loss and gain as the river flows downstream until the end of the fan, and finally Toyohira-gawa joins the river Ishikari-gawa. On the other hand, waters are dominantly lost from the riverbed except at the proximal fan in the high-water season. The ratio of river runoff between low-water season and high-water season at the datum discharge observation station of the river (Kariki

Figure 11.   Curves showing both observation and calculation results of the river runoff at Moiwa Station, Sapporo.

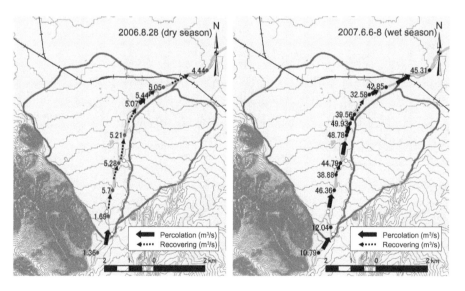

Figure 12.   Amount of surface runoff and the differences in recharge and discharge among the riverbed segments in dry and wet season.

Bridge Station) are approximately 1 to 10. Thus, active interaction between river water and groundwater in Toyohira-gawa is occurring through the year. But, there is no difference that can be seen on the relationship between river runoff and water gained/lost from the riverbed during these 11 years (1995–2006). For example, both gained water i.e. maximum discharge from groundwater and lost water i.e. maximum recharge to groundwater are $0.85 \, \mathrm{m^3/s}$ and $0.63 \, \mathrm{m^3/s}$, while river runoff are $25.28 \, \mathrm{m^3/s}$ and $23.24 \, \mathrm{m^3/s}$ at that time.

3.3 *The planar distribution of groundwater potential*

Groundwater levels in the fan usually change due to the snow melting season in spring (Apr. to Jun.) and the rainy season in autumn (Sep. to Oct.), while water shortage seasons are in summer and winter (Fig. 13). As shown in Fig. 14, contours of the groundwater table are similar to the topographical surface of the fan. The main direction of the groundwater flow is estimated from the proximal to distal fan for the watershed region at the central part of Sapporo surface. However, at the Mid-fan the groundwater flow is dissociated from the Toyohira-gawa and flows to the outer directions of the fan. There is about 10 m topographical difference in height between the Sapporo surface and the Hiragishi surface, and the groundwater table is interrupted at the intersection of these two surfaces. On the other hand, as shown in Fig. 14, the recharge mechanism of water from the River to the underground is dominated at the mid-fan area. Contours are disturbed at some places on the fan due to groundwater abstraction. The Figure also shows difference of groundwater levels of around 5m between the water-rich and water-deficient seasons.

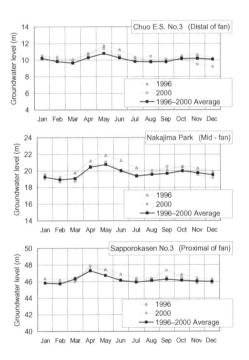

Figure 13.   Seasonal variation of groundwater levels.
Note: High groundwater levels in April to May and low groundwater levels in February to March.

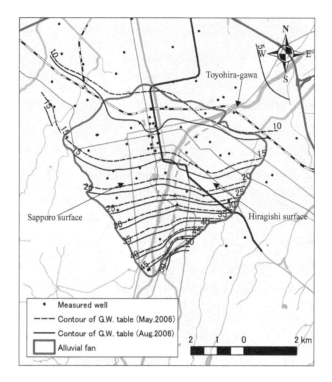

Figure 14.   Planer potential distribution map of groundwater (May and Aug. 2006).

3.4   *Distribution of profiled potential of groundwater*

The equipotential map shows that the recharge from the proximal to mid-fan successively percolates into aquifer II to III, II to IV and with increasing depth in aquifer III (Fig. 15). Records of the groundwater potentials, observed in sporadically distributed observation wells on the fan indicate a downward ground-water flux in the different aquifers and/or in the same aquifer. However, the result of groundwater quality analyses show that it flows from the proximal to distal of the fan. Also, the results of ^{14}C dating analyses of groundwater show that younger surface water becomes older as it percolates into the deeper level. From these facts, it appears that precipitation and river water permeate from the proximal to mid-fan surface to the underground, and then it flows down to the distal fan.

3.5   *Interaction of surface water and groundwater*

The water rise of the River is observed twice, once during the snow melting season in spring and the other during the rainfall season in autumn, whereas the shortage of water is experienced during summer and winter. Thus, maximum river runoff is in May, and minimum in August (Fig. 11). The difference between maximum and minimum river water volume is measured as $100\,m^3/s$ to $10\,m^3/s$. It is obvious that there are water gaining segments and water losing segments on the river throughout the year, and/or water gaining or losing segments are frequently changing either seasonally or by the fluctuation of the precipitations (Fig. 16).

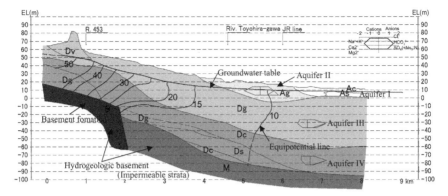

Figure 15. Profiled groundwater potential map of the fan.
Abbreviations: A; Holocene, D; Pleistocene, B; Tertiary bedrock, M; Hydrological basement strata (mud stone), v; pyroclastics (volcanic ash etc.), s; sand, g; gravel, c; silt to clayey. The typical hexagonal diagrams in each aquifer are also presented in the Fig (see text).

Date of observation	Gorin Bridge	Moiwa Bredge	Befor join place with Yamahana-gawa	S22jou Bredge	Horohira Bridge	Minami Bridge	Ichijo Bridge	Mizuho Bridge	Azuma Bridge	Kariki Bridge / Observation Station (Kariki)	
24.Aug.95	1.42	0.37	0.70	0.42	-1.04	-0.09	2.38	-1.11	-0.78	-1.15	①
12.Oct.95	0.87	0.19	-0.16	-0.32	-1.52	-0.04	0.65	0.43	-0.23	-0.55	②
19.Dec.95	1.89	0.38	-0.66	0.76	1.51	0.16	-0.99	-1.18	-1.11	-2.35	③
2.Aug.96	1.74	-0.70	0.05	-1.27	-0.75	-1.33	1.84	1.08	-2.98	0.57	④
12.Aug.96	1.26	-0.10	0.49	-1.36	-0.79	-0.10	0.73	-0.13	-0.59	-0.96	⑤
5.Sep.96	1.85	-0.66	0.12	-1.47	-0.52	0.29	0.48	1.63	-2.17	0.35	⑥
27.Sep.96	1.46	-0.21	1.84	-1.83	-0.45	-4.37	4.17	0.87	-6.32	-1.33	⑦
20.Oct.04		0.48	-0.83	2.29	-1.50	-0.51	-0.08	-0.72	1.49	-0.49	⑧
28.Oct.04		0.78	-1.02	2.37	-2.29	-0.56	-0.10	-0.41	1.53	-0.20	⑨
11.Nov.04		0.87	-1.02	2.57	-2.42	-0.58	0.30	-0.58	1.10	-2.83	⑩
28.Aug.06	0.04	0.09	-0.17	-0.28	-0.15	-0.14	0.37	-0.39	-0.61	-1.66	⑪
12.Oct.06		0.48	-0.83	2.29	-1.50	-0.51	-0.08	-0.72	1.49	-0.49	⑫

Figure 16. Figure shows the losing water and gaining water segments in the riverbed at each observation dates. ①–⑫; Observation dates of runoff, numbers are common to Fig. 17. ///-0.10 Segment of losing water and its volume. █ 1.00 Segment of gaining water and its volume (m³/s). Numbers represent the discharge volume of the surface water (m³/s), values with negative signs imply a decrease while those with positive signs imply increase of discharge volume in the respective segments.

River runoff of Toyohira-gawa has a strong correlation with precipitation as shown in Fig. 17. River runoff increases with precipitation, e.g., in case of ①1st~31st Aug. 1995, ⑦ 10th Sep.~20th Oct. 1996 and ⑫ 20th Sep.~20th Oct. 2006, and all but one (③21st Sep~19th Oct. 1995) of the remaining seven cases also show a slight correlation.

Increases in river runoff or precipitation correspond to the increases in groundwater level, especially when precipitation is greater than 25 mm/d (①, ②, ③, ⑦ , ⑫, in Fig. 17). However, in case of precipitation less than 25 mm/d, there is no relationship between runoff and groundwater levels even though some correlations for precipitation and river runoff exist.

There is a positive correlation between the increasing river runoff and groundwater level rise for precipitations of less than 25 mm/d, in spring, as 20 days before the increase in river runoff due to snow melting, the groundwater levels abruptly rise (Fig. 18).

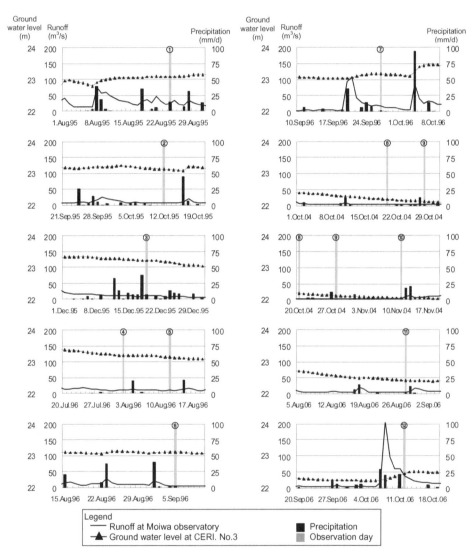

Figure 17.   Correlation between precipitation, runoff and groundwater levels. ①–⑫: Observation dates of runoff. Values in each figure are recorded after observation for 30 days. Numbers of ① to ⑫ are common to Fig. 16.

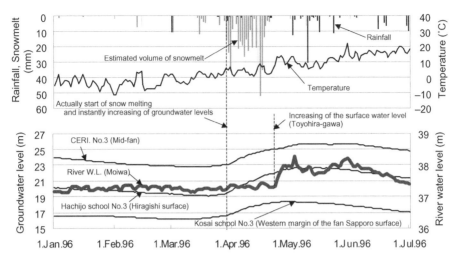

Figure 18. Comparison of river water level and groundwater level in the snow melt season.
Note that groundwater levels rise during the actual snow melting time, however, river water level increases even three weeks after the snow melting. Location of three observation wells is shown in Fig. 19.

Although, it is obvious that the cause of increase of river runoff is snow melting, it is not clear what exactly causes the groundwater level rise. However, the reason may be as follows.

Mean annual precipitation around the Toyohira-gawa alluvial fan area is 1,162 mm, and of it 416 mm (36%) is snowfall (Mukai et al., 2008a). Mean annual temperature in Sapporo is 7.8°C (1901–2000), but during snow melting season in spring, the mean monthly temperature in March is 0°C and maximum temperature is 4°C during last 30 years, while the mean monthly temperature of April is 8.5°C. From these facts, it seems that if the coefficient of snowmelt is 3[1] (this value matches with the snow melt timing in the Sapporo area), groundwater levels rise due to the infiltration into the underground from the melting at the base of the snow. Detailed examinations have been done on the land-use map of the Toyohira-gawa alluvial fan (Fig. 19) that shows that the ratio of non-concrete covered surface to concrete covered surface, which are occupied by buildings and pavements, is nearly 35:65, and that this will have enough permeability to allow infiltration into the ground from the bottom of the snow. But the knowledge of the differences in permeability for both the asphalt concrete-covered surfaces and non-concrete-covered surface is limited. Uncertainties also arise due to the lack of knowledge of the detailed geologic and hydro-geologic structure of the surface of the fan, relationships between groundwater levels and infiltration, and presence of water-saturated areas, among others. It is difficult to address the problems here, but recharge from the snow bottom on the fan surface at snow melting season will be of great possibility to raise the groundwater levels.

---

[1] The reproduction method for snowfall and snowmelt is supplied by Sugawara (1972). Snowmelt quality, coefficient K, and the water equivalent of snowmelt from the bottom of the snow m_B, were adjusted (to $K = 3$ and $m_B = 0$) to give good reproduction of snow depth change.

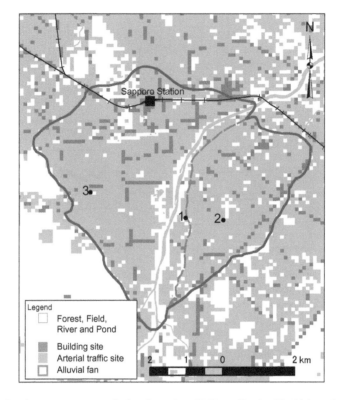

Figure 19. Land use map around the fan. 1; CERI well, 2; Hachi-jo school well, 3; Kosai school well.

3.6   *Water balance of the fan*

Simulation of the fluctuation of groundwater levels was carried out by using the SHER model (**S**imilar **H**ydrologic **E**lement **R**esponse **M**odel, 2002) by assuming a fixed rate of the River runoff from 1st Jan. 1996 to 31st March, 2000. Meteorological data (temperature and precipitation) was sourced from the Sapporo meteorological observatory. The result shows that the total amount of rainfall, snow melt and leakage from the main water distribution systems is 1,241 mm/y, and 226 mm/y of surface water is lost as evapo-transpiration and 638 mm/y by runoff and that the remaining 277 mm/y is recharged to the ground. 598 mm/y is recharged to the aquifer (of which 321 mm/y is recharged from the riverbed and the remaining 277 mm/y from the surface) and abstraction is 329 mm/y. So, discharge from the aquifers will be calculated as 197 mm/y and only 72 mm/y will be stored in the aquifer every year (Mukai et al., 2008b).

4   CONCLUSION

The interaction between surface water and groundwater in the Toyohira-gawa alluvial fan needs discussion. Water gaining segments and water losing segments in the river through the year are recognized, and it is obvious that in the case of over 25 mm/d of precipitation, an immediate infiltration from riverbed to the groundwater occurs. Moreover, recharge occurs from the bottom of the snow due to its melting because the groundwater levels always rise before the rise of river runoff in the spring season. Thus, river runoff of Toyohira-gawa depend not only precipitation but also on discharges from the ground-water. Koizumi et al. (2008) calculated its volume to be 6.24 million m^3/y. The same phenomena were found and reported at Echi-gawa alluvial fan area, Shiga Pref., central Japan (Kobayashi et al., 2008,2009) and Shigenobu-gawa alluvial fan area, Ehime Pref., Shikoku Islands (Wata-Nabe et al., 2008, 2009).

Even though six various alluvial fans are picked up for detailed hydro-geological study in Japan by **R**esearch group on **H**ydro-environments around alluvial **F**ans in Japan, only the Toyohira-gawa fan is used as an urban city. Over 600 thousand out of 1.9 million populations of the fan area lived in the urban center of Sapporo city and not only the central and local government offices are established on the fan, but also the most activities of the city life is based on the fan. So the Toyohira-gawa became important for the city life for its water supply, sewage disposal use, water utilization and drainage of water, including for Sustainable Urban Drainage Systems (SUDS). Today, 6.3 m^3/s of water is supplied to Sapporo city. Surface water supply is in excess of 8.6 m^3/s, even during the drought years. Because of this reason, it is believed that for Sapporo city and its adjacent areas (total population of 2.15 million) there is little need to use the groundwater (Hu, 2008b).

It is currently estimated that there are three hundred million cubic metres of groundwater in the Toyohira-gawa alluvial fan basin, and about 20 million cubic metres of water is abstracted each year (Hu, 2008a). Hence, there are sufficient amounts for current use, in terms of both surface water and groundwater. However, the climate is continuously changing all over the world and it is difficult to anticipate sudden sharp reductions of precipitation or unbalanced rainfall. Such situations might seriously reduce surface water availability and lead to a rapid increase of groundwater use. According to simulation results for the future climatic conditions, the maximum pattern of precipitation will be changed from spring to autumn by the end of this century (Koizumi et al., 2008).

From these considerations, it would be important to build common awareness of both surface water and groundwater.

ACKNOWLEDGEMENTS

The authors would like to put forward their heartfelt thanks to Drs. Kayane Isamu (Prof. Emeritus of Tsukuba Univ.), Kuroki Mikio (Hokkaido Univ.), Sakura Yasuo (Chiba Univ.), Kobayashi Masao (Osaka Kyoiku Univ.) and Shimada Jun (Kumamoto Univ.), committee members of RHF, for their continuous encouragement, constructive guidance and valuable suggestions at different stages of our present survey and research work. Thanks are also due to staffs of Hokkaido Regional Development Bureau of MLIT, Ishikari River Development and Construction Department and River Management Office of Sapporo, HKD for their help and logistic supports in different form. We are thankful to all the members of RHF for their helpful advice and critical discussions during to study around the fans in Japan. The authors would also like to express their appreciation to Mrs. Hashimoto Nana of Raax Co. Ltd. for her assistance in preparation of Figures.

REFERENCES

Daimaru, H. (1989) Holocene Evolution of the Toyohira River Alluvial Fan and Distal Floodplain, Hokkaido, Japan. *Geographical review of Japan*, 62A-8, 589–603 (in Japanese with English abstract).

Hu, S.G. (2008a) River Toyohira-gawa Alluvial Fan (General Remark). RHF (ed.) Hydro-environments of Alluvial Fans in Japan. *Monograph of 36th IAH Congress*, IAH 2008 Toyama, 1–12.

Hu, S.G. (2008b) Water Resources of Sapporo- Present and its Future. RHF (ed.) Hydro-environments of Alluvial Fans in Japan. *Monograph of 36th IAH Congress*, IAH 2008 Toyama, 79–94.

Ichimaru, H., Kayaki, T., Watanabe, O., Hijii,T., Saito, M., Yanagida, M. (2008) Groundwater Flow System in Shigenobu-gawa Alluvial Fan. RHF (ed.) Hydro-environments of Alluvial Fans in Japan. *Monograph of 36th IAH Congress*, IAH 2008 Toyama, 283–294.

Ikeda, M., Takata, S., Matsueda, H. (1998) Estimated Value of the Environmental Tritium Concentration and the Altitude Isotope Effects of δD and $\delta^{18}O$ in Hokkaido. Radioisotopes, 47: 812–823.

Kayaki, T., Watanabe, O., Ichimaru, H., Hijii, T., Saito, M., Yanagida, M. (2008) Water budget of Shigenobu-gawa Alluvial Fan-Particularly the Interaction Between its Surface Water and Groundwater. RHF (ed.) Hydro-environments of Alluvial Fans in Japan. *Monograph of 36th IAH Congress*, IAH 2008 Toyama, 295–304.

Kobayashi, M., Hijii, T., Yan, H. J., Hamada, T. (2008) Relations Between Groundwater and River Water in the Echi-gawa Alluvial Fan. RHF (ed.) Hydro-environments of Alluvial Fans in Japan. *Monograph of 36th IAH Congress*, IAH 2008 Toyama, 223–234.

Kobayashi, M., Hijii, T., Yan, H.J., Hamada, T. (2009) Study of groundwater Flow System in the Echi-gawa Alluvial Fan. Printed in this book.

Nagaoka, D., Koizumi, K., Mukai, K., Hu, S.G. (2008) Geomorphological Development and Hydrogeology in the Tohyohira-gawa Alluvial Fan, Central Hokkaido, Japan. RHF (ed.) Hydro-environments of Alluvial Fans in Japan. *Monograph of 36th IAH Congress*, IAH 2008 Toyama, 13–28.

Mukai, K., Konishi, H., Koizumi, K., Hu, S.G. (2008a) On the Groundwater Flow system around Toyohira-gawa Alluvial fan. RHF (ed.) Hydro-environments of Alluvial Fans in Japan. *Monograph of 36th IAH Congress*, IAH 2008 Toyama, 29–44.

Mukai, K., Koizumi, K., Nagaoka, D., Hu, S.G. (2008b) On the Present State of the Water Balance of Toyohira-gawa Alluvial Fan. RHF (ed.) Hydro-environments of Alluvial Fans in Japan. *Monograph of 36th IAH Congress*, IAH 2008 Toyama, 45–62.

Koizumi, K., Mukai, K., Konishi, H. and Hu, S.G. (2008) Change of Water Balance by Global Warming in the Toyohira-gawa Alluvial Fan. RHF (ed.) Hydro-environments of Alluvial Fans in Japan. *Monograph of 36th IAH Congress*, IAH 2008 Toyama, 63–78.

SHER Model (2002) http://www.arsit.or.jp/koukaiyou/sher2.htm   http://www.arsit.or.jp/koukaiyou/manual.PDF

Sugawara, M. (1972) Method of Run-off Analysis, Kyoritsu publication, 257 p (In Japanese).

Takahashi, K., Hu, S.G., Jimbo, M., Hayama, H., Tanaka, Y. (2003) Character of the underground water in the Riv.Bibi-gawa lowland from the radio isotopic point of view. Groundwater of the Shikotsu Volcanic Ash Plateau, Hokkaido, Japan-Formation, Resources, Circulations, Quality, Recharges and its Circumstances—Suzuki et al (ed.), *Monograph of IAHS/AISH*, Sapporo, 47–60.

Watanabe, O., Kayaki, T., Ichimaru, H., Hijii T. (2009) The groundwater flow system characterized by hydrogeologic structures and the mechanism of interaction with surface water at Shigenobu-gawa alluvial fan, Ehime Pref., Japan. Printed in this book.

CHAPTER 13

Present state of the water balance in the Isawa-gawa alluvial fan, Iwate Prefecture, Japan, and its future prospect under global warming

Kazutaka Yoshimatsu, Hiroshi Nakura, Kiichiro Sato &
Toshiki Kobayashi
Fukken Gijyutsu Consultant Co., Ltd., Aoba-ku, Sendai, Japan

ABSTRACT: The authors carried out a series of studies of the water balance in the Isawa-gawa alluvial fan, Iwate Prefecture, Northeast Japan. The study focuses on generating basic data that can help towards the sustainable use of groundwater as well as the general management of both surface water and groundwater. Specifically, studies have been carried out to (a) prepare a model of the topographical features of this fan, (b) understand the water balance between groundwater in the widespread aquifer of the fan and the rivers Isawa-gawa and Kitakami-gawa, and (c) speculate on the prospect of the future water balance between surface water and groundwater. As a result, the authors understood the hydrogeomorphological and hydrogeological outline of the Isawa-gawa alluvial fan. Hydraulic simulations suggest that the effect of global warming on the water balance of the Isawa-gawa alluvial fan will be relatively small, although the level and storage of groundwater (recharge—discharge) will fall from spring to autumn, whilst the increases of precipitation and thawing will increase them from autumn to spring.

Keywords: Isawa-gawa alluvial fan, cylindrical water division, inclined multi-staged terrace, global warming

1 INTRODUCTION

The Isawa-gawa alluvial fan, also called the Isawa Plain, has an area of about 110 km^2 and is one of the largest fans in Japan (Fig. 1, Fig. 2). People in this area have made use of the very gentle fan topography as paddy fields, although the topographic feature sometimes caused struggles for water.

The Isawa-gawa alluvial fan is topographically characterized as a northward 'inclined fan' or 'inclined stepped terrace', with the water of the Isawa-gawa River, which runs along the northern margin of the fan, hard to make use of. To ease the water shortage problem, many small aqueducts (e.g. Juan-zeki and Shigeira-zeki aqueducts) were built and engineering heritages such as cylindrical water diversion works still remain. The water shortage problem was accelerated by the relatively low precipitation in this region: i.e. 1,100–1,800 mm per year, compared with the mean annual precipitation of 1,700 mm in Japan.

The purposes of this project are (1) to know the outline of the geomorphological features, (2) to analyze the hydrological relationship (water balance) between the extensive aquifer

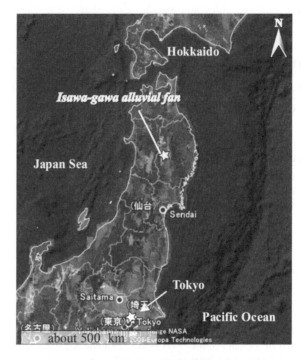

Figure 1.   Location map of the Isawa-gawa alluvial fan (Google Earth).

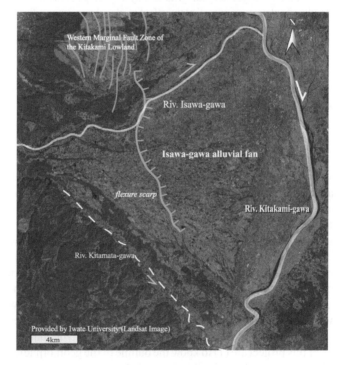

Figure 2.   Aerial photograph of the Isawa-gawa alluvial fan.

of the fan and the Isawa-gawa–Kitakami-gawa river system, and (3) to make a hydrological prediction for the global warming period in the near future. The authors hope that this project will provide basic data for the future proper use of the groundwater of the fan. The authors also hope that this project will be of some use to the synthetic management of groundwater and surface running water, i.e. the aquifer of the Isawa-gawa alluvial fan and the river water of the Isawa-gawa–Kitakami-gawa river system.

2 OUTLINE OF THE ISAWA-GAWA ALLUVIAL FAN

2.1 *Topography and geology*

The Isawa-gawa alluvial fan has the area of about 110 km^2 and is one of the largest fans in Japan.

The fan surface is in fact an aggregation of stepped terrace surfaces that started to form in Middle Pleistocene time, with the highest surface lying 80–100 m above the present river bed. The Isawa-gawa River presently runs along the northern margin of the fan, and the alluvial plain (flood plain) is only narrowly distributed along the river channel.

Small thicknesses of alluvial fan deposits also characterize the Isawa-gawa river fan. In particular the thickness of the fan deposits on old terraces near the fan head is only several meters. Previous studies thus proposed that the Isawa-gawa alluvial fan is an erosional fan or terrace topography (Saito, 1983; Watanabe, 1991).

The drainage area upstream of the head of the Isawa-gawa alluvial fan is only 190 km^2. However the area is the source of abundant debris, because the area is occupied mainly by Neogene (Miocene) sedimentary rocks such as tuff and sandstone, and Quaternary volcanic products that tend to cause landslide and slope failure on steep mountainsides.

2.2 *Hydrological environment*

The Isawa-gawa and Kitamata-gawa rivers run along the northern and southern margins of the Isawa-gawa alluvial fan, respectively, eroding mountains and hills. These rivers cannot supply surface running water and groundwater to the Isawa-gawa alluvial fan, because the surface of the fan is an aggregate of terrace surfaces and stands higher than the river beds. Hence the annual precipitation of 1300 mm (the AMeDAS data from the Wakayanagi observational site on the Isawa-gawa alluvial fan) accounts for most of the natural water supply to the fan.

In addition to the natural water supply, settlers have built up some systems of water supply on the Isawa-gawa alluvial fan since the 17th century or earlier. For example, many irrigation canals from small upstream artificial water reservoirs feed water to the paddy fields on the terraces of the fan.

Moreover old topographic maps show that there were many irrigation ponds on higher terrace surfaces and around the fan head. Although some of them were designed just to collect natural rainfall, most of them were formed at the head of a small marsh on a terrace where surface water and infiltrated water tend to gush out and accumulate. Old settlers seem to have made use of the high-level underflow in the terraces due to the small thickness of the terrace-forming sand and gravel.

3   GEOMORPHOLOGICAL ANALYSES

The Isawa-gawa alluvial fan is one of the largest fans in Japan, having the longitudinal dimension (East–West, from head to fringe) of about 18 km and lateral dimension (North–South, between fringes) of about 23 km. Terrace topography is well developed on the fan with the maximum relative height between the highest terrace surface and the present river bed of 100 m.

Moreover some of the western marginal faults of the Kitakami Lowland, marking the boundary between the Ou Mountains and the Kitakami Lowland, seem to cut the Isawa-gawa alluvial fan; the presence of flexure scarp around the midfan has been suggested.

Under these circumstances, we collected information in the following ways to understand the regional features of the geomorphological structure and, if possible, hydrological environment of the Isawa-gawa alluvial fan.

- Landform division and analysis of geomorphological structure from aerial photograph interpretation

  We interpreted grayscale 1:40,000 scale aerial photographs, taken in 2000 and downloadable from a Geographical Survey Institute website, "land vicissitude archive · aerophotograph browse system." The authors made a landform division (terrace surface division) of the Isawa-gawa alluvial fan, marked tectonic features such as fault and flexure scarps, and interpreted their relationships with the terrace surfaces.
- Analysis of hydrological environment

  Irrigation canals on the Isawa-gawa alluvial fan from the Ishibuchi Dam (completed in 1953) in the upstream of the river and from other old small aqueducts have been well prepared and maintained in recent years. They supply agricultural water to the paddy fields which are widespread on the Isawa-gawa alluvial fan. However, before the preparation and maintenance of these irrigation canals, the main sources of water supply to the fan were small aqueducts of the 16th century or older, spring water on the fan surface, running water in a small stream, and rainfall.

  To understand the natural hydrological environment of the fan without the effects of artificial water supply, we marked irrigation ponds, stream landforms, and downstream directions on old topographic maps published in 1915.

From these assessments based on two pre-existing sources of information, the authors listed hydrogeomorphological features of the Isawa-gawa alluvial fan and made a landform division to construct a hydraulic model.

4   ANALYTICAL RESULTS

The result of landform analysis from aerial photographs and characteristic features of lateral sections are shown in Fig. 3.

A segment of the western marginal faults of the Kitakami Lowland runs just around the midfan, and the terrace surfaces younger than the M1 surface have been displaced and form a flexure scarp. On the other hand, the higher terrace surfaces along the southern extension of the fault segment have not clearly been displaced, suggesting that the fault segment has been inactive there.

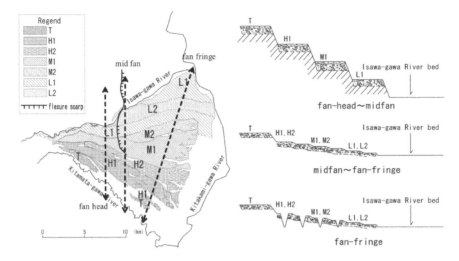

Figure 3.   Result of landform analysis and characteristic features of lateral cross sections of fan-head to fan fringe areas.

Table 1.   Characteristics of each terrace surface.

Landform division	Relative-height from the present river-bed	Thickness of sediments		Age
		Tephra	Gravel bed	
L2	3~10 m	–	?	~10 ka
L1	5~25 m	–	10 m>	10~20 ka
M2	10~25 m	2 m	?	80 ka?
M1	15~40 m	2~3 m	10 m>	80~120 ka
H2	30~40 m	2~3 m	3~10 m>	150~300 ka
H1	60~80 m	3 m	3~10 m>	
T	80~100 m	5~6 m	3~10 m>	200~400 ka

Relative height from the present river bed is the largest at the fan head, the smallest around the midfan, and a little larger in the fan fringe.

We referred to the descriptions of Watanabe (1991) for the thickness of the terrace sediments. The sediments on all surfaces are generally more than 10 m thick at the fan head, whereas in the fan fringe the thickness of the gravel bed on higher terrace surfaces is several meters or so.

We referred to Koike et al. eds. (2005) for the formation age of each terrace surface, taking new results of tephrochronology into account.

On the upstream side of the flexure scarp, the level difference between terrace surfaces is large and terrace scarps are evident. On the downstream side, by contrast, the level difference between terrace surfaces is low and planation process has obscured the terrace boundaries. These features are important points in examining the hydraulic structure.

Previous studies of the landform division of the Isawa-gawa alluvial fan have also clarified the thickness and facies of the sediments on each terrace surface, and have made clear description of the tephra beds on the surface (e.g. Watanabe, 1991). Table 1 shows the thickness and age of terrace sediments described in the previous studies. The authors present a

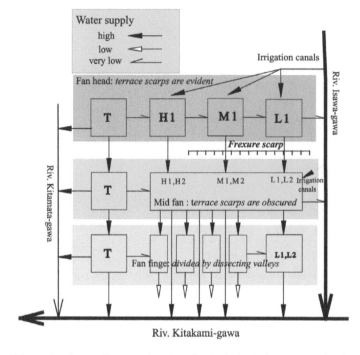

Figure 4.   Schematic planar diagram showing the hydrological structure in the Isawa-gawa alluvial fan.

new landform division based on the landform analysis from aerial photographs, change of topographic features in longitudinal and lateral directions, and hydrological environment analyzed from old topographic maps. In addition to the traditional terrace surface division, we discriminated geomorphological changes in longitudinal direction and divided the fan into three areas with specific lateral topographic profile.

As shown in Fig. 3, the three areas in the Isawa-gawa alluvial fan can be characterized as follows: (1) fan-head–midfan (flexure scarp) area with marked terrace scarps and a larger level difference between terrace surfaces, (2) midfan (flexure scarp)–fan-fringe area with obscure terrace scarps and a smaller level difference between terrace surfaces, and (3) fan fringe area divided by longitudinal dissecting valleys. We constructed a hydraulic model of the Isawa-gawa river fan on the basis of the new landform division shown in the previous section, and considering the fact that the water on the terrace surfaces basically runs in the maximum dip (longitudinal) direction (Fig. 4).

5   RECENT METEOROLOGICAL CONDITIONS

The authors collected meteorological data to analyze the recent meteorological condition of the Isawa-gawa alluvial fan and its vicinity. The observational sites and collected data are as follows.

[Wakayanagi]:1979–2007 (for 29 years)
[Ishibuchi Dam]:1985–2004 (for 20 years)

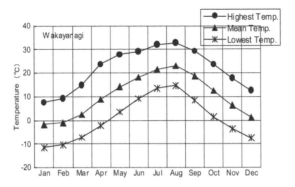

Figure 5.   Monthly highest, mean, and lowest temperatures at the Wakayanagi observation site.

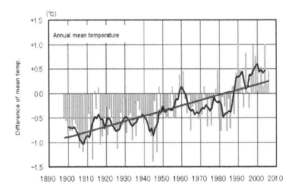

Figure 6.   Fluctuation of the annual mean temperature of Japan from 1898 to 2006 (Japan Meteorological Agency, 2007).

5.1   *Temperature*

Fig. 5 shows the monthly mean temperature of Wakayanagi. At Wakayanagi, the mean temperature of August, 23.2°C, marks the annual highest monthly mean temperature, and that of January, −1.7°C, marks the lowest. Fig. 6 shows the fluctuation of the annual mean temperature of Japan during the recent 100 years, with the average increase being 1.1°C/100 years (0.01°C/y). Fig. 7 shows the fluctuation in the annual mean temperature at Wakayanagi. At the site, temperature has been increasing a little more rapidly than its average increase in Japan. However, since Fig. 7 shows the temperature fluctuation for a different time range from Fig. 6, it is not proper to say that the warming in the Isawa-gawa alluvial fan and its vicinity is particularly advancing. Comparing each element in Fig. 7, the increase in the minimum temperature is particularly large.

5.2   *Precipitation*

The mean annual precipitation of the Wakayanagi, and Ishibuchi Dam sites are 1,301 mm, and 1,790 mm, respectively, indicating that the precipitation increases toward the fan head (mountain area). Compared to the mean annual precipitation of Japan of 1,700 mm, the precipitation on the Isawa-gawa river fan is less.

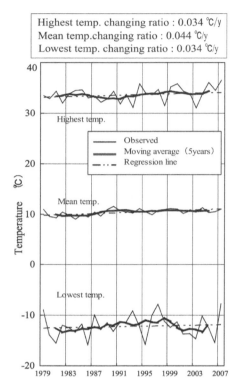

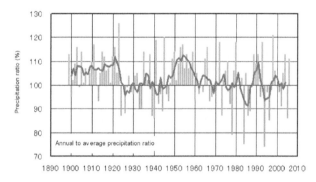

Figure 7.   Fluctuation of annual highest, mean, and lowest temperature at the Wakayanagi observation site.

Figure 8.   Fluctuation of annual to average precipitation ratio of Japan (bars) from 1898 (Japan Meteorological Agency, 2007).

The annual precipitation of each site rises to the maximum in August and falls to a minimum in February. Fig. 8 shows the long-term fluctuation in the annual to average precipitation ratio for Japan from 1898 to 2006 (Japan Meteorological Agency, 2007). Fig. 8 implies that the annual precipitation of Japan is slowly decreasing, and the precipitation difference between dry years and wet years has been increasing for the recent 20–30 years.

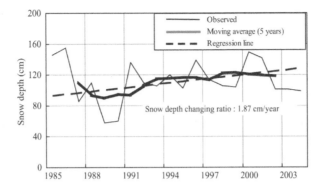

Figure 9.   Fluctuation of annual maximum snow depth at the Ishibuchi Dam site.

5.3 *Evapotranspiration*

The annual evapotranspiration of Wakayanagi (annual precipitation = 1,301 mm) is 662 mm, indicating that more than half of the annual precipitation is lost as evapotranspiration. In other words, another half of the precipitation becomes surface running water or groundwater recharge.

Since the monthly evapotranspiration calculated with the Thornthwaite method is a function of the monthly mean temperature, it rises to a maximum in August, when the monthly mean temperature is maximum, and becomes zero in January and February, when the monthly mean temperature is below zero.

5.4 *Snow depth*

Fig. 9 shows the temporal fluctuation of the annual maximum snow depth at the Ishibuchi Dam site. The mean annual maximum snow depth is 112 cm, whereas the maximum depth from 1985 to 2004 was 155 cm and minimum depth was 58 cm. The annual maximum snow depth has been increasing by 1.87 cm/y in average.

6   STREAM DISCHARGE OF THE KITAKAMI-GAWA RIVER

The groundwater and stream water discharged from the Isawa-gawa River and Isawa-gawa alluvial fan run into the Kitakami-gawa River along the fringe of the fan. To know the annual fluctuation of the flow of the Kitakami-gawa River, its stream discharge was measured.

Peaks in stream discharge occur in April and August. The peak in April is likely caused by snow melt whereas that in August is by precipitation.

The precipitation and discharge in August are well correlated, suggesting that the stream discharge of the Kitakami-gawa River at this time has an intimate relationship to the precipitation.

The correlation between the maximum snow depth and the discharge in April is, on the other hand, not so good. It may be because the amount of snowmelt is not only a function of snow depth but also a function of temperature.

7   GROUNDWATER

7.1   *Distribution of wells*

In the Isawa-gawa alluvial fan there are a number of wells, which were used from 1985 to 1987 for groundwater level observation (Iwate Construction Work Office, Tohoku Regional Bureau, Ministry of Construction, 1988). There were 55 observation sites, all of which were artesian wells in the gardens of private houses. The observation wells were distributed across the whole fan, although the number of the wells gradually decreases towards the fan head. The depth of the wells was mostly less than 10 m.

7.2   *Fluctuation of groundwater level*

The groundwater level is observed to lie in the terrace deposits or in the alluvium on the river beds, i.e. close to the topographic surface (Iwate Construction Office, Tohoku Regional Bureau, Ministry of Construction, 1988). The groundwater level fluctuation in a well in the paddy fields (Fig. 10) shows that the groundwater level tends to start rising at the onset of the irrigation period (from May to August) and to start falling as soon as the irrigation period ends. The data suggest that the infiltrated water from the paddy fields plays an important role as a recharge source of the groundwater.

7.3   *Distribution of groundwater*

The observation data indicate that the water level fluctuation in most wells is less than several meters, and there is no correlation between landform division and groundwater level fluctuation. Moreover, if we examine the groundwater level fluctuation of three laterally divided areas, i.e. the fan head, midfan, and fan fringe areas, there is no significant difference.

Thus we conclude that the thickness of the terrace deposits is almost even all over the fan and that the water table in the fan lies in the terrace deposits and is close to the topographic surface.

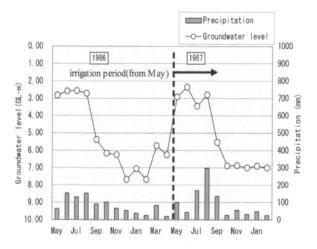

Figure 10.   Fluctuation of ground water level compared with precipitation.

8 NUMERICAL SIMULATION OF WATER BALANCE

8.1 *Area of numerical simulation and boundary conditions*

The area of analysis of the model was the Isawa-gawa alluvial fan, bound on north by the Isawa-gawa River, east by the Kitakami-gawa River, and south by the ridge line on the hill along the Kitamata-gawa River. The model domain was about 18 km from east to west, and 23 km form north to south. Using a uniform cell size of 100 m, this model area was discretized for the whole domain. The top surface of the model was the topographic surface and the base represented the bottom of the aquifer (base of the Kazawa formation).

The condition on the boundaries is as follows. The Isawa-gawa River and Kitakami-gawa River are set as known hydraulic head boundaries and the water levels of the rivers were given as the fixed water levels. The southern boundary ridge was considered as a no-flow boundary, because the ridge coincides with an underground watershed.

Further, a drainage effect was set to every surface point to estimate the surface runoff and groundwater discharge. The area of numerical simulations and boundary conditions are shown in Fig. 11.

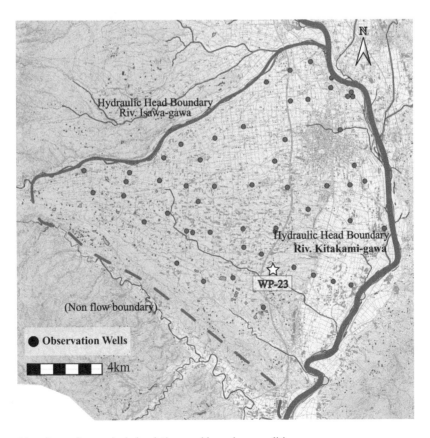

Figure 11. Area of numerical simulations and boundary conditions.

170 *Kazutaka Yoshimatsu et al.*

8.2 *Aquifer and hydraulic parameters*

As aquifer layers, the alluvium on the present river bed (Kitakami-gawa River deposits), terrace deposits, and the Kazawa Formation were selected. The terrace deposits were further grouped in their respective terrace levels. The hydraulic parameters of each aquifer layer were decided from trial calculations with the aid of preexisting data (Iwate Construction Work Office, Tohoku Regional Bureau, Ministry of Construction, 1988). Figs. 12 and 13 show the plan and cross sections, respectively of the numerical model, and Fig.12 presents the hydraulic parameters of each aquifer layer decided from the trial calculations.

8.3 *Setting up the groundwater recharge*

The authors divided the area of analysis into paddy fields and the other areas (Fig. 14) because the authors considered, from the observation of groundwater levels, that the infiltrated water from paddy fields is likely to have a large influence on the groundwater levels. Further the authors divided the period of analysis into two for the groundwater recharge: the irrigation period when paddy fields are in use, and the non-irrigation period when paddy fields are not in use.

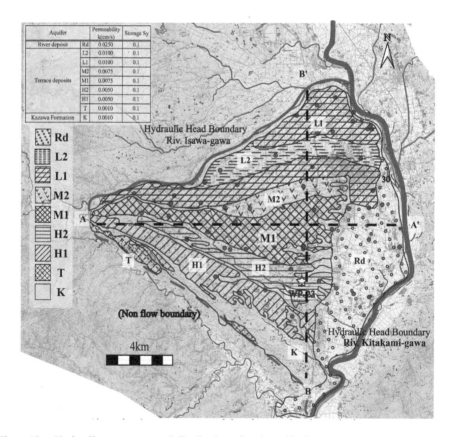

Figure 12.   Hydraulic parameters and distribution of each aquifer layers.

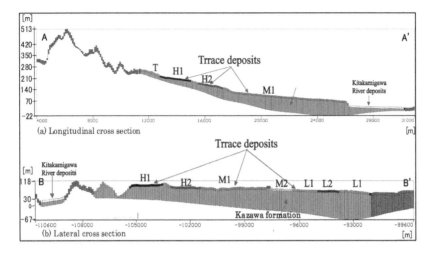

(a) Longitudinal cross section

(b) Lateral cross section

Figure 13.   (a) Longitudinal cross section (in Fig. 12 A-A′ section) and (b) lateral cross section (in Fig. 12 B-B′ section), of the numerical simulation model.

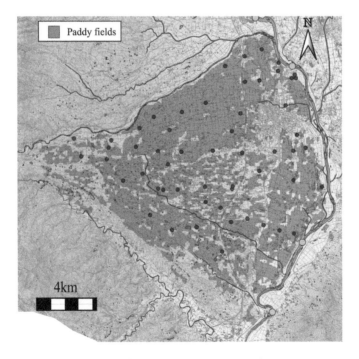

Figure 14.   Paddy fields in the area of numerical simulation model.

8.3.1   *[Irrigation period] May 1 to September 7*

Paddy fields: The amount of infiltrated water from paddy fields was set to be a part of the water requirement; i.e. the *groundwater recharge* was defined as the difference between the water requirement and the evapotranspiration. The water requirement was set to be

7 mm/day from trial calculations based on the increase of the groundwater level in the irrigation period.

The other areas: The difference between the precipitation and evapotranspiration was set to be the *groundwater recharge*.

8.3.2   *[Non-irrigation period] September 8 to April 30*

Everywhere, the difference between the precipitation and evapotranspiration was set to be the *groundwater recharge*. The snowfall in winter, however, was treated as follows.

Daily mean temperature ≤ 0°C: The precipitation was treated as snowfall and was added to the snow depth. No *groundwater recharge* was calculated.

0°C < daily mean temperature ≤ 2°C: The precipitation was treated as snowfall and was added to the snow depth. The difference between the snowmelt and evapotranspiration was set as the *groundwater recharge*. The snowmelt was subtracted from the snow depth. The snowmelt was calculated, using the day of disappearance of snow depth, to be 4 mm/day.

2°C < daily mean temperature: The precipitation was treated as rainfall. If there remains snow depth, snowmelt was added to rainfall, and the difference between the calculated rainfall and evapotranspiration was set as the *groundwater recharge*.

8.4   *Numerical simulation method and results in reproductive model*

The authors performed three-dimensional transient analysis using the finite-difference method. The lattice interval for the finite-difference method was 100 m in the whole area of analysis. The period of analysis was from 1985 to 1987, and the timestep of calculation was 5 days. The code selected for simulation was MODFLOW-2000; a modular, three-dimensional, finite difference, groundwater flow model developed by the U.S. Geological Survey (2000). The application for pre- and post-processing in this study was Visual MODFLOW (Waterloo Hydrogeologic, Inc.).

The calculation results were verified with the observed groundwater level during the period of analysis (Fig. 17). The correlation between the observed and reproduced groundwater levels is fairly good over the whole area of analysis (Fig. 15), and the authors

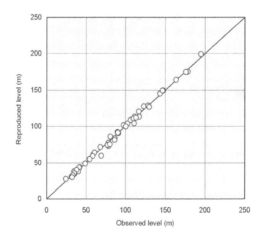

Figure 15.   Correlation of observed and reproduced ground water levels.

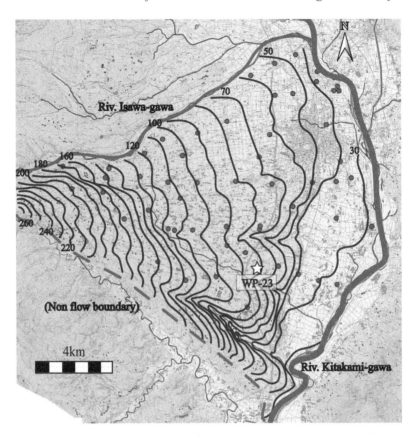

Figure 16. Contour diagram showing the configuration of the water table calculated with the reproductive model.

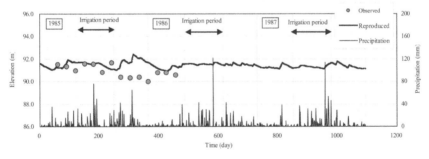

Figure 17. Fluctuation of ground water levels comparing the observed (WP-23) and reproduced values.

hence evaluate that the level and configuration of the water table are mostly reproduced (Fig. 16).

The annual mean of the total discharge in the analytical period was approximately 430,000 m^3/day, of which the surface runoff was 390,000 m^3/day, comprising 91%. The remaining 9% was the groundwater discharge of 40,000 m^3/day. Since the surface runoff also finally runs into the Kitakami-gawa River, the total water supply from the Isawa-gawa alluvial fan to the Kitakami-gawa River corresponds to the above total discharge. On the other hand, the annual mean discharge of the Kitakami-gawa River at the Kozenji observation site, which is downstream of the Isawa-gawa alluvial fan, is approximately 26,400,000 m^3/day. Hence the contribution ratio of the water supply from the Isawa-gawa alluvial fan, as given by the total of surface runoff and groundwater discharge, to the discharge of the Kitakami-gawa River is 1.6%.

9   GLOBAL WARMING PROJECTION MODEL

The authors used a regional climate model (RCM20) developed by the Japan Meteorological Agency for the global warming projection in and around the Isawa-gawa alluvial fan. Using the RCM20 model, Japan Meteorological Agency performed the reproduction calculation of the climate from 1981 to 2000, and it further made the climate change projection from 2081 to 2100 using the A2 emissions scenario of the IPCC Special Report on Emissions Scenarios. As the representative meteorological data, we adopted the results of the calculation point that is closest to the Wakayanagi observational site on the Isawa-gawa river fan, because the results of the calculations are given in the resolution of 20-km mesh.

Meteorological observation site Wakayanagi N39°7.9', E141°3.8'
RCM20 calculation point N39°8.1', E140°57.3'

9.1   *Comparison of RCM20 reproduction values with predictions*

Monthly mean temperatures are predicted to increase, with a predicted increase in mean annual temperature from 1981–2000 to 2081–2100 of 2.8°C. The temperature increase in April and from October to December are particularly large.

The monthly mean precipitation from January to July tends to decrease in the future, whereas that from August to October tends to increase. The annual precipitation is predicted to increase by 105 mm, although the simulated baseline values of precipitation from July to September are significantly larger than the observations.

The decrease in snowfall is one of the predicted climate changes associated with the increase in mean temperature, particularly in winter time. Our prediction of monthly mean temperature, adding the amount of calculated temperature change of each month to the base data measured at the Wakayanagi observational site, suggests that the mean temperature for all months will not go down below zero, taking the observations of 1982 as the base data. Moreover the snow cover period in winter will be short, because the first snowfall will come later and the thaw will set in earlier. The shortening of snow cover period indicates that the *groundwater recharge* in winter will increase.

On the other hand, annual evapotranspiration will increase by 100 mm (15%), at a constant rate from spring to autumn.

9.2   *Conditions of numerical simulations*

The authors calculated the *groundwater recharge* in the same way as for the model under current (or baseline) conditions. This time we used the predicted meteorological parameters as calculated in the previous section. The authors put the newly calculated projection of the *groundwater recharge* in the model and calculated the future water balance. The authors

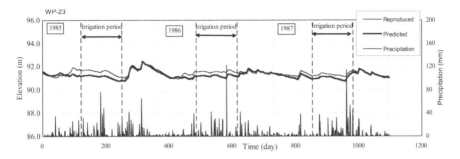

Figure 18. Fluctuation of ground water levels at (WP-23) comparing the predicted and reproduced values.

suggest that the influence of global warming is reflected in the difference of the predicted water balance from the current water balance.

Boundary conditions and hydraulic parameters are the same as those in the baseline model.

9.3 *Results of numerical simulation in the prediction model*

Fig. 18 compares the reproduced and predicted groundwater levels to show the changes in groundwater level.

There is little difference between predicted (future) and reproduced (baseline) ground-water levels, as shown by the change of groundwater level of between -21 cm to $+15$ cm. Considering the seasonal change in the groundwater level, the level tends to fall from April to September and rise up from October to March. Thus the authors interpret that the decrease in precipitation and increase in evapotranspiration from spring to autumn cause the decrease in the *groundwater recharge*, whereas the increases in precipitation and snow melt from autumn to spring cause the increase in the *groundwater recharge*.

The authors input the amount of water supply (precipitation and water requirement of paddy fields) as the element of recharge and calculated the surface runoff and groundwater discharge to rivers, but the two discharge elements do not change largely. Our calculation virtually shows no change in the groundwater discharge to the rivers, suggesting that the influence of global warming on the water supply to the Kitakami-gawa River will be very small.

These hydraulic simulations suggest that the effect of global warming on the water balance of the Isawa-gawa alluvial fan is relatively small, although the level and storage of groundwater (recharge—discharge) will fall from spring to autumn, and the increases of precipitation and thaw will allow levels to rise from autumn to spring.

10 MEASURES TO PREVENT THE FALL OF GROUNDWATER LEVEL

Our simulation of the influence of global warming on the water balance predicted that groundwater levels will fall from spring to autumn. The Isawa-gawa alluvial fan is widely occupied by paddy fields and much agricultural water is used in the irrigation period of

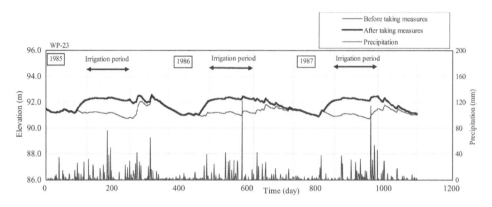

Figure 19.   Fluctuation of future groundwater levels before and after taking measures.

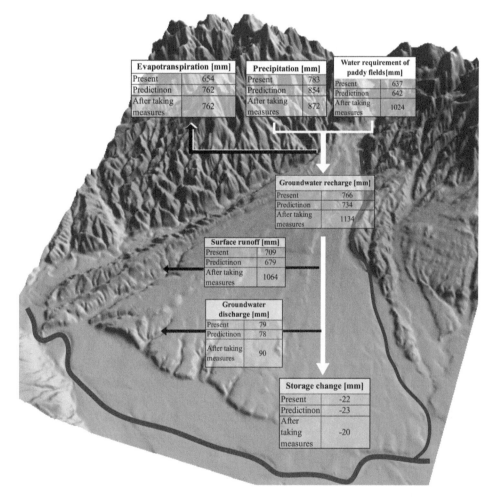

Figure 20.   Annual simulated water balance of the Isawa-gawa alluvial fan for present, predicted future, and future after taking measures.

early spring. Moreover there are many shallow wells. The fall in groundwater level may have adverse impacts on these water supplies. On the other hand, global warming will shorten the snow cover period and increase the snowmelt discharge. Hence the authors expect that the decline in groundwater levels can prevented by starting the irrigation in April or earlier, when surface water flows are increased by snow melt. The authors will verify the effectiveness of the measures using an analytical model. The authors carried out another simulation for the case where the irrigation period is from April and the water requirement of the paddy fields is 10 mm/day. Fig. 19 shows an example of the fluctuations of the groundwater level after taking these measures.

The groundwater level will start to rise at the onset of the irrigation period and becomes almost constant from May onwards. After the end of the irrigation period, the levels will decline until they reach the same level as before the measures. Each element of recharge and discharge will increase at the onset of irrigation, stay constant afterwards, and finally attain the original level after the end of the irrigation period. The groundwater storage before taking measures will tend to decrease in the irrigation period, but it will tend to increase even in the irrigation period after taking measures. Thus it is verified that advancing the irrigation period and increasing the water requirement of the paddy fields will effectively prevent the decrease in groundwater level.

11   CONCLUSION

The authors understood the hydrogeomorphological and hydrogeological outline of the Isawa-gawa alluvial fan from available data and materials, inferred the state of groundwater flow, and constructed a hydraulic model that can reproduce the present state. With this model, the authors predicted the hydrological effects of global warming.

As the result, the global warming was predicted to have very little effect on the water balance of the fan as well as the Kitakami-gawa River. This is because the storage capacity of the aquifer in the Isawa-gawa alluvial fan, including the subsurface distribution of the Kazawa Formation, is so large that small variations in precipitation do not significantly change the discharge from the fan (Fig. 20).

Although the total effect of global warming will likely be small, our analysis predicts that the groundwater level under the fan will tend to reduce from spring to autumn. Hence the authors carried out another simulation in the case that the timing of irrigation is one month earlier than usual and the water requirement of paddy fields increases. According to the simulation, the groundwater level will tend to start rising at the onset of irrigation period and then decline to the original level soon after the end of the irrigation period. The simulation suggests that the fluctuation of the water requirement of the paddy fields has a great influence on the groundwater level.

ACKNOWLEDGMENT

We would like to appreciate various instructions and cooperation during the course of this project. We are particularly indebted to Mr. Hiroyuki Seki, the former Head of Iwate Office for River and National Highway, Tohoku Regional Bureau, Ministry of Land, Infrastructure, Transport and Tourism for his precious advice and guidance in founding the Alluvial Fan

Research Group. We also would like to thank Iwate Office for River and National Highway for the permissions to use materials. Thanks are extended to Mr. Shigeki Kanoh, the former Head of Construction Office of Isawa Dam, for his attendance to the First Committee and his on-site explanation of the Isawa Dam and the Isawa plain. They worked as very precious helps to the foundation of the research group. We thank Dr. Ian Holman, the co-editor of this volume, and two anonymous reviewers for their critical reading of the manuscript. We finally thank to the members of Iwate Branch, Fukken Gijyutsu Consultant, Co., Ltd. for their dedication to the First Committee.

REFERENCES

(Former) Iwate Construction Work Office, Tohoku Regional Bureau, Ministry of Construction (1988): 1987 outsourcing report of the water-balance analysis in the Isawa and Senmaya areas. (in Japanese)

Japan Meteorological Agency (2005): Global warming projection, Volume 6.

Japan Meteorological Agency (2007): Climate change monitoring report 2006. 25 p. (in Japanese)

Koike, K., Tamura, T., Chinzei, K., Miyagi, T., eds. (2005): Topography of Japan, 3. Tohoku Region. University of Tokyo Press, 355 p. (in Japanese)

Saito, K. (1988): The alluvial fans of Japan. Kokon Shoin, Tokyo, 280 p. (in Japanese)

Watanabe, M. (1991): Chronology of the fluvial terrace surfaces in the Kitakami Lowland, Northeast Japan, and fluctuations of debris supply during the Late Pleistocene. The Quaternary Research (Daiyonki Kenkyu), 30: 19–42. (in Japanese with English abstract)

CHAPTER 14

Study of the groundwater flow system in the Echi-gawa alluvial fan, Shiga Prefecture, Japan

Masao Kobayashi
Department of Natural Sciences, Osaka Kyoiku University, Kashiwara Osaka, Japan

Makoto Yamada
Institute of Geothermal Sciences, Kyoto University, Noguchibaru, Beppu, Oita, Japan

Heejun Yang
Department of Natural Sciences, Osaka Kyoiku University, Kashiwara Osaka, Japan

Takaaki Hijii
Geo Field Survey Consultant, Higasiku Fukuoka, Japan

ABSTRACT: To understand the groundwater flow system in the alluvial fan of the River Echi-gawa (-gawa means river, here after Echi-gawa or the River), we conducted a hydrogeological survey with emphasis on the observations of the groundwater level, and sampling of groundwater and surface water for the analysis of water quality and stable and radioactive isotopes. Groundwater in the shallow aquifer originates from precipitation and irrigation water (river water or abstracted groundwater) which penetrates into the deeper layers and, except in the area at the centre of the fan on the left bank of the Echi-gawa where there is an area of flowing artesian conditions, generally flows in a north-west direction. Shallow groundwater also discharges into the river in the area around the apex of the fan and is recharged by river water in the area around the edge of the fan on the left bank, and flows in two directions near the residual hill in the downstream area. Groundwater in the shallow and intermediate aquifers in the fan actively circulates and mixes with water flowing vertically or laterally in the groundwater system. On the other hand, groundwater in the deep aquifer is recharged by groundwater around the intermountain area and flows towards the downstream area at an extremely low speed and is almost static. It is also noted that excessive groundwater abstraction from deep wells greatly affects the water level in the deep aquifer in the alluvial fan of the Echi-gawa.

Keywords: Alluvial fan, groundwater flow, water quality, stable and radioactive isotopes

1 INTRODUCTION

The Echi-gawa alluvial fan is a complex fan which is located on the east bank of Lake Biwa, Shiga Prefecture (Fig. 1). This area has the highest utilization ratio of groundwater in the entire prefecture. Referring to the groundwater in this region, there are some hydro-geological studies (Takaya & Nishida, 1964), regarding groundwater utilization in the aquifer (Motokage & Minami, 1975; Horino et al., 1989; Horino & Maruyama, 1990 etc.).

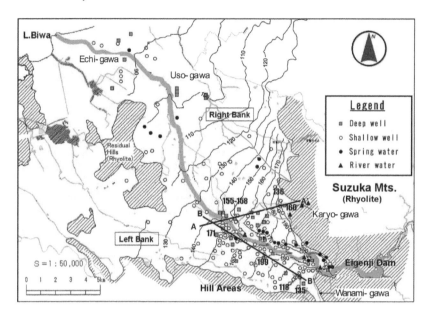

Figure 1.   Study area and the observation sites of the shallow and deep wells and sampling of waters. Transect lines A–A' and B–B' refer to the geological cross-sections in Figure 5. The numerical values along the transect lines show the station numbers of the inspection wells.

Horino et al. (1989) reported that irrigation water losses to the aquifer are 6.4–8.4 mm per day, indicating that irrigation water contributes greatly to groundwater recharge. However, there are very few surveys available of the groundwater flow in the upstream area of the fan, particularly in the medium and deep-seated aquifers down to 150 m depth below ground level (G.L.) and many aspects of the relationship between the groundwater and surface water are yet to be studied in this region. To understand the groundwater flow system in this complex alluvial fan, we conducted a hydrogeological survey with an emphasis on the observation of groundwater levels, and the sampling of groundwater and surface water for the analysis of water quality and stable and radioactive isotopes in the alluvial fan of the Echi-gawa and the surrounding area, during the period from 2004 to 2008. This paper presents the main findings of the survey on the groundwater flow system in the fan.

2   SUMMARY OF THE GEOGRAPHICAL AND GEOLOGICAL CHARACTERS OF THE STUDY AREA

The Echi-gawa alluvial fan is formed by the rivers Echi-gawa, Wanami-gawa and Karyo-gawa. Echi-gawa is the major river in the drainage basin of Lake Biwa, having a length of 50 km and catchment of about 200 km^2. The fan spreads to the northwest from the mouth of the valley (at an elevation of approximately 200 m above mean sea level) of the Echi-gawa, and 3 terraces (higher, middle and lower) have developed along the river. The higher terrace, however, does not exist on the left side of the Echi-gawa. In the lowland area, below an elevation of about 100 m, a number of natural levees and back marshes can be observed around the river channel, and also several residual hills are present (Fig. 1). Further downstream, the delta is formed broadly in the lowland area near the lake (Uemura, 1979).

The basement of the study area is mainly underlain by Koto Rhyolites, which consist of granitic rocks and rhyolites, and the Ko-Biwako Group deposits in the upper part of the basement. In this region, the upper Ko-Biwako Group (Gamo Formation), which covers the Ko-Biwako Group, consists of a few fine-grained sand/gravel beds and a muddy stratum (Shiga Pref. 1979). These layers are repeatedly stratified and there is little continuity between each layer, particularly the 5–10 m thick silt and clay layers which are sandwiched in between. The aquifer (down to 150 m below ground level) is classified into 3 types from their hydro-geological characteristics, i.e. free groundwater at shallow level (unconfined aquifer) and confined groundwater at the medium and deeper levels. However, in the downstream section of the river, the groundwater in the shallow aquifer is confined. Further details of the geomorphology and geology of the study area are described in Oishi et al. (2008).

3   METHOD

Groundwater levels of the shallow wells (up to 30 m deep) and deep wells (down to between 30 and 150 m depth) were measured once or 4 times per month, except for the inspection wells where the water level was measured using automatic self-recorders, and single water samples of groundwater, river water, spring water, rice paddy water, and rain water were collected in August at 238 stations in total (Fig. 1). Groundwater samples from the deep wells were collected separately from 1–6 layers consisting of sand and gravel using a bailer. Water temperature, pH, and electrical conductivity (E.C.) were measured directly at the well site using a mercury thermometer, pH colorimeter, and E.C. meter, respectively. Major chemical constituents were analyzed with an ion-chromatograph, and also pH4.8 alkalinity was obtained by titration with 0.02 N H_2SO_4. The analyses of stable ($\delta^{18}O$, δD) and radioactive (Tr. and ^{14}C) isotopes were performed at the Laboratory of Geological and Nuclear Sciences, New Zealand, and the detection limits are $\pm0.1‰$ for $\delta^{18}O$, $\pm1‰$ (δD), ±0.1T.U.(Tr) and $\pm1{\sim}4‰$ (^{14}C), respectively. Radon concentrations were measured by the analytical method (Horiuchi & Murakami, 1977) and the detection limit is 0.1 Beq/l.

4   GROUNDWATER FLOW TRACED BY THE GROUNDWATER POTENTIALS

4.1   *Seasonal change of groundwater level*

Fig. 2 shows examples of the seasonal changes of the groundwater levels of shallow and deep wells through the year from January 2007 to April 2008. Groundwater levels of shallow wells show seasonal changes in the fan, being high in the irrigation period and rainy seasons, and low in the non-irrigation period (particularly in the winter season). This indicates that precipitation and irrigation water affect the groundwater levels of the shallow aquifer (wells) in the fan. These results are consistent with that of Horino et al. (1989), i.e., irrigation water greatly contributes to the groundwater recharge described above.

On the other hand, deep wells of depths of between approximately 30–120 m showed two observable patterns. One shows the same seasonal change in water levels as shallow

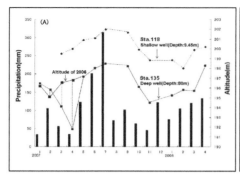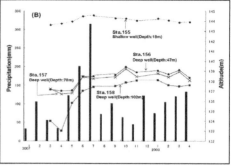

Figure 2.   Seasonal changes in groundwater levels of the shallow and deep wells in the fan. Figure (A) is the area at the apex of the fan on the left bank of the Echi-gawa (Stas. 118, 135). Figure (B) is the area at the edge of the fan on the right bank (Stas. 155–158).

wells and is observed in the area at the apex of the fan (Fig. 2(A)) and at the center of the fan. This implies that the groundwater in the deep aquifer is likely to be semi-confined, and groundwater in the upper layer permeates into the deeper aquifers in these areas.

Another pattern is observed at the edge of the fan (Fig. 2(B)). Groundwater levels of deep wells in the area at the edge of the fan are low during March to April, but remain high and nearly constant during other months. These seasonal changes of water level in deep wells do not show clear correlation with the fluctuation in precipitation and the irrigation period.

This indicates that the precipitation and irrigation waters does not directly affect the deep groundwater. Note that the water levels of deep wells (Stas. 135) showed significant decreases from January to April 2007, because the precipitation (snow in winter season) was exceptionally low and a large quantities of water was pumped from deep wells for the agriculture sector during the period.

4.2   *Annual variation ranges of groundwater levels in the shallow and deep wells*

Fig. 3 shows an example of isopleths of the annual fluctuation ranges (hereafter the fluctuation ranges) in the shallow and deep wells. Here, the fluctuation ranges shown are the maximum range during the years from 2004 to 2007.

The shallow wells on the low terrace plain of the Echi-gawa fan showed a small fluctuation range of less than 3 m (Fig. 3(A), except for two areas which showed large fluctuation ranges exceeding 5 m (maximum 8 m) at the edge of the fan. Whereas, the fluctuation ranges are generally of approximately 4 m at the center of the Echi-gawa fan on the left side (Higashi-Omi city) of the Echi-gawa (hereafter Left Bank) and at the edge of the Karyo-gawa fan on the right side (Echi Country) of the Echi-gawa (hereafter Right Bank). On the Right Bank, the fluctuation ranges of shallow wells on the terrace plain of the Echi-gawa were generally smaller as compared to the Left Bank.

On the other hand, some deep wells on the Left Bank showed large fluctuation ranges exceeding 11 m; however, their fluctuation ranges are generally similar to those in the shallow wells, i.e., ranging 1–4 m. On the Right Bank, however, two concentric areas

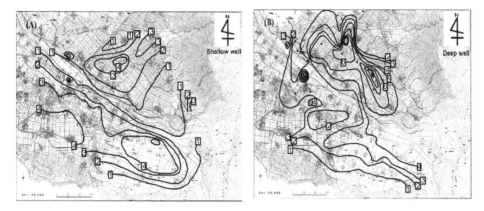

Figure 3.   Isopleths of the range (m) of annual changes of groundwater levels in the shallow (A) and deep (B) wells.

showed large fluctuation ranges exceeding more than 6 m (maximum 10 m); the isograms of these areas were completely different from those of the Left Bank.

The large geographical difference in the fluctuation ranges of the shallow and deep wells mentioned above is related to the distribution of the wells from which large quantities of water are abstracted. According to the survey conducted by Higashi-Omi City (2007) on the effect of groundwater pumping on groundwater level in 2005, the continuous pumping of 1,000 m^3 of water per day from shallow wells (which caused the water level to decrease by 12.4 m) affects shallow wells (shallow aquifer) to a radius of about 350 m, but does not significantly affect deep wells. Similarly, the continuous pumping of 10,000 m^3 of water per day (causing the water level to decrease by 5.5 m) affects deep wells to a radius of more than 400 m, but does not affect shallow wells.

As described above, the effect of water pumping from a well on the groundwater level varies with the depth of the well. However, it is clear that the effect is significant in the case of both shallow and deep wells. The magnitude of the effect of groundwater pumping on the groundwater level differs for shallow groundwater and deep groundwater; this difference can be presumably explained by the difference in the geological structures of this area.

In particular, a thick clay layer (thickness of more than approximately 10 m) exists in the area where the effect of pumping water from deep wells on groundwater levels is small in the shallow wells but great in the deep wells. We presume that the thick clay layer divides shallow groundwater and deep groundwater, resulting in the difference in the effect of groundwater pumping on groundwater level.

4.3   *Spatial distribution of contours of groundwater levels*

The contour map of shallow groundwater levels in Fig. 4 is similar in form to that of the land inclination of the fan (cf. Fig. 1), and the flow direction of the shallow groundwater around the area of the apex of the fan is generally towards the river of the Echi-gawa, indicating that shallow groundwater in this area recharges the river water.

However, on the Left Bank, the flow paths which are running in a north–west direction along the old river channel in the centre of the plain diverge towards two directions near

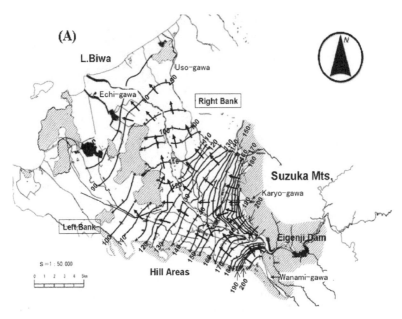

Figure 4.   Contour map of groundwater levels in the shallow aquifer around the Echi-gawa alluvial fan on October 2007. Arrows indicate the inferred flow lines of shallow groundwater.

the residual hill site in the downstream area (Fig. 1). Although the results are not shown in the figure, deep groundwater contours show a pattern similar to the shallow groundwater. However, on the Right Bank, the groundwater level around the edge of Karyo-gawa fan is lower than that of the neighbouring area, indicating groundwater in the deep aquifer flows toward the area where there are some deep wells for industrial and water supply use where a large amount of groundwater is abstracted.

4.4   *Vertical distribution of groundwater potential*

The water levels in the shallow groundwater in the fan are always higher by 3–5 m or more than those of the groundwater in the deep aquifers as shown in figure 2, indicating a descending flux of groundwater and the permeation of groundwater into the lower layers is expected in most areas of the fan.

As shown in Fig. 5, the configurations in each cross section of the equi-potential lines are different between the Right Bank (A–A′ cross-section) and Left Bank (B–B′). The equi-potential lines generally show a sharp upward convex curve facing the mountain in the shallow aquifer, and are vertical in the layers of the middle and deep aquifers on the Right Bank and have a gentle upward convex curve facing the mountain or are vertical on the Left Bank, except the area at the center of the fan (Sta.199) where a flowing well is located. These lines around the center of the fan (Sta.199) show an upward convex shape.

These results indicate that the shallow groundwater percolates into the deep layers vertically and the water recharges at the mountains and flows in a downstream direction through the deeper layer in the alluvial plain on the Right Bank. And, on the Left Bank, the groundwater generally flows in a downstream direction parallel to the land surface, giving evidence

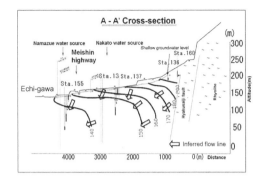

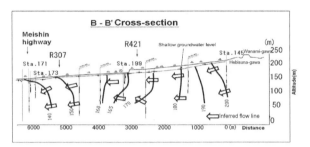

Figure 5.   Vertical distributions of groundwater potentials in the fan on March 2007. Arrows indicate the inferred flow lines of groundwater.

of the upward flow of the groundwater at the centre of the fan as shown by the flow lines in the figures.

5   GROUNDWATER FLOW TRACED BY WATER QUALITY OF GROUNDWATER AND SURFACE WATER

5.1   *Spatial distribution of water quality of groundwater and river water*

As shown by the Stiff diagrams in Fig. 6, most of the water samples were found to be Ca-HCO$_3$ type but differences in their compositions (Na-HCO$_3$ and Ca-SO$_4$ types) are dependent on location- the Na-HCO$_3$ and Ca-SO$_4$ (spring water) type of water are influenced by the chemical characteristics of the type of bedrock (rhyolite is rich in Na and SiO$_2$) and by fertilizer (NH$_4$SO$_4$) dissolved under the paddy field (Kobayashi et al., 2007), respectively.

The values of electrical conductivity (E.C.) of shallow groundwater normally ranged from 0.1 to 0.20 mS/cm, and were generally high on the terrace plain of the Karyo-gawa fan on the Right Bank, and low (below 0.1 mS/cm) around the river channel of the Echi-gawa and the intermountain area (Fig. 7). Also, most of the low measured SiO$_2$ concentrations of up to 20 mg/l were distributed in the piedmont area of the fan on the Right Bank and around the residual hills in the lowlands on the Left Bank (Fig. 8). In particular, the values of E.C. and SiO$_2$ concentrations of the shallow groundwater near the river channel are comparatively low, and similar to those of river water in the downstream area from the edge of the fan on

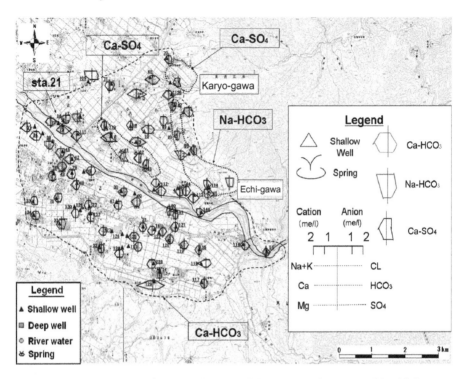

Figure 6.   Spatial distribution of water quality type of all water samples shown with Stiff diagram.

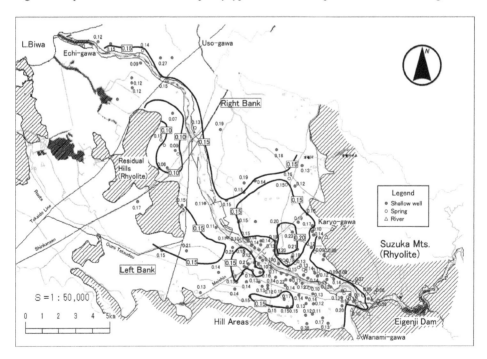

Figure 7.   Isopleths of average electrical conductivities (mS/cm) of shallow groundwater.

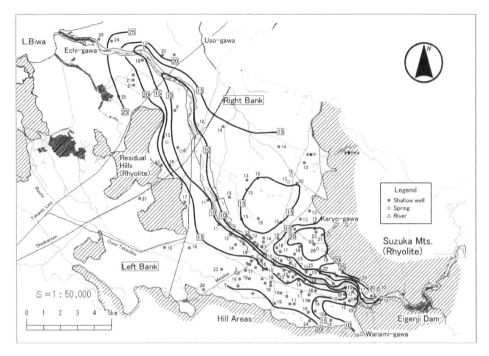

Figure 8.   Isopleths of average SiO₂ concentrations (mg/l) of shallow groundwater.

the Left Bank. Further the isopleths of E.C. and SiO₂ concentrations match comparatively well with the flow direction of shallow groundwater in this area (Fig. 4). It seems to be that river water (river-bed water) percolates into the shallow aquifer near the channel.

5.2   *Vertical distribution of water quality of groundwater*

Figs. 9 and 10 show examples of the vertical distribution of water quality of groundwater in the aquifer together with a geological cross section along the transect lines of A–A′ on the Right Bank and B–B′ on the Left Bank and the isopleths of E.C. and SiO₂ concentrations are also shown in the figure. Chemical characteristics of groundwater on the Right Bank generally showed a small vertical change, although the concentrations are different from site to site, i.e., each of the E.C. and SiO₂ concentrations of groundwater at the area of the apex of the fan is low in E.C. and high in SiO₂, whereas this is reversed in the area at the border of the fan. However, the water quality of the groundwater is different in character in each of the shallow, medium and deep-seated (depths exceeding 80–100 m) aquifers. For example, SiO₂ concentrations of the groundwater range from 10~20 mg/l in the shallow aquifer to 20~30 mg/l in the medium aquifer, and are over 30 mg/l in the deep-seated aquifer.

The water quality in the deep-seated aquifer on the Right Bank is generally similar to that at the intermountain area, indicating that the groundwater in the deep-seated aquifer is recharged by seepage water from the water body in the mountain or river waters from the drainage basins of the Karyo-gawa and tributaries.

On the other hand, on the Left Bank, the vertical profile of water quality of these components differs from that on the Right Bank. This is because of a thick permeable layer

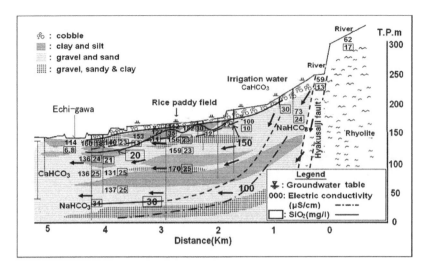

Figure 9.   Vertical distribution of water quality of groundwater along transect of A–A' on the Right Bank. Each of dotted an thick lines in the figure show the isopleths of E.C and SiO₂ concentrations. Arrows show the inferred flow lines of groundwater estimated by the results of water quality.

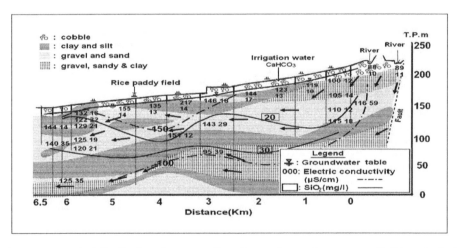

Figure 10.   Vertical distribution of water quality of groundwater along transect B–B' on the Left Bank. Dashed and thick lines show the isopleths of E.C and SiO₂ concentrations, respectively. Arrows show the inferred flow lines of groundwater estimated by the results of water quality.

which is found above the deep aquifer around the area of the apex of the fan, and a thick clay (confining layer) found in the deep level layer in the area at the center of the fan as shown in the figure. Characteristically, the aquifers in the fan are roughly classified into three types of water quality, i.e., shallow level aquifer having high electrical conductivity, which has a good correlation with Total Dissolved Solids but not with the colloidal SiO₂, the medium having medium ones, and the deep-seated aquifer having generally low E.C. and high SiO₂ concentrations. Generally, the features of the isopleths of the water quality in the aquifers

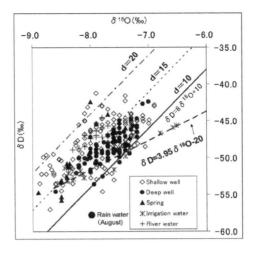

Figure 11.   Delta diagram of all samples of groundwater and the waters of river, paddy field, spring and rain. Thick line ($d = 10$) indicates the meteorological water line.

match the geological structures (shown in Figs. 10 and 11). These results indicate that the groundwater in the aquifer flows according to the geological structure in the fan.

6   GROUNDWATER FLOW TRACED BY THE CHARACTERISTICS OF STABLE AND RADIOACTIVE ISOTOPES OF GROUNDWATER AND SURFACE WATER

6.1   *Spatial distribution of the $\delta^{18}O$ values of groundwater and surface water*

Stable isotope ratios of groundwater collected from the shallow and deep wells, river water, spring water, and paddy field water fluctuate between approximately −7.0‰∼−9.0‰ for $\delta^{18}O$ and −40‰∼−60‰ for δD. As the $\delta^{18}O$ of all water samples plot on or around the meteorological water line in the delta diagram (Fig. 11), the characteristics of the isotopic compositions of the water samples will be discussed using the $\delta^{18}O$ value.

Fig. 12 shows an example of the distribution of the $\delta^{18}O$ values of shallow groundwater in the non-irrigation period. Generally, the $\delta^{18}O$ values of shallow groundwater in the area around the top of the fan and the piedmont area are small, less than −8.0‰, and increase downstream. However, high values (−7.2∼−7.4‰) were obtained at the edge of Karyo-gawa fan on the Right Bank and the center of the alluvial plain on the Left Bank. The reason is that, in these areas, irrigation water having high concentrations of $\delta^{18}O$ (approx. −6.5∼−7.5‰) percolates into the shallow aquifer and mixes with groundwater flowing from the upper region. However, the mixing ratio of the $\delta^{18}O$ values between irrigation water and groundwater could not be determined yet.

The result is not consistent with that of the vertical distributions of groundwater potentials in the aquifer in the fan. On the other hand, the $\delta^{18}O$ values of the shallow groundwater near the river channel were comparatively small (−7.6∼−8.0‰), and the isopleths of the values generally match well with that of the flow lines of shallow groundwater in this area (Fig. 4). The feature of the isopleths of the $\delta^{18}O$ values for the period is similar to that

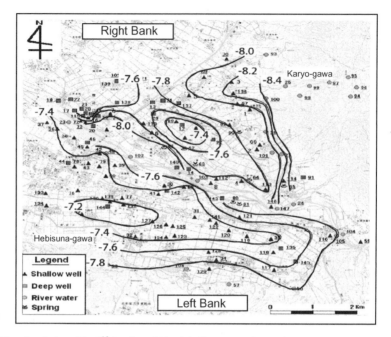

Figure 12.   Isopleths of the $\delta^{18}O$ values (‰) of shallow groundwater in the non-irrigation season.

in an irrigation season, although there is a little difference in the values ($-0.2\sim-0.4$‰). These results indicate river water (river-bed water) seepage into the shallow aquifer near the channel because the $\delta^{18}O$ values of river water did not show noticeable seasonal changes (-8.0 ± 0.2‰) at every site.

6.2   *Vertical distribution of the $\delta^{18}O$ values of groundwater*

The vertical change of the $\delta^{18}O$ values of groundwater in the aquifer is generally small, as with water quality. However, there is a difference of approximately -1.0‰ between the shallow aquifer and the deeper aquifer (Figs. 13 and 14).

Although the $\delta^{18}O$ values of groundwater in the deep aquifer are lacking at some sites, groundwater in the aquifer is characteristically divided into three zones as shown in the figures. However the depth of each zone is different from site to site. The reason is that the geological condition of the layer on the Left Bank is different from that on Right Bank, i.e., both on the Left Bank and Right Bank thick gravel and sand layers are distributed in the area of the apex of the fan. However, on the Left Bank, a thick clay layer is widely inter-bedded in the deep layer, while on the Right Bank, comparatively thin clay and sandy layers with little continuity between each layer are repeatedly stratified as shown in the diagram. Furthermore, the $\delta^{18}O$ in the deep level groundwater showed small (light) values ranging from approximately -8.0 to -8.5‰ and matched the values obtained from groundwater and river water at both areas around the piedmont and the apex of the fan.

Thus the water in these deeper layers is assumed to flow from a deep flow system that, for the right bank, is recharged by seepage water from a groundwater body in the mountain or

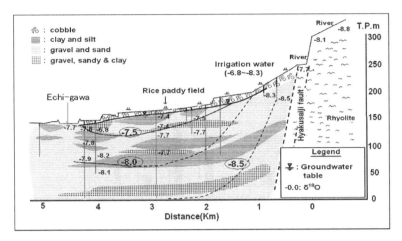

Figure 13.   Vertical distributions of $\delta^{18}O$ values (‰) of groundwater along the transect line A–A' on the Right Bank. Thick line shows the isopleths of $\delta^{18}O$ value.

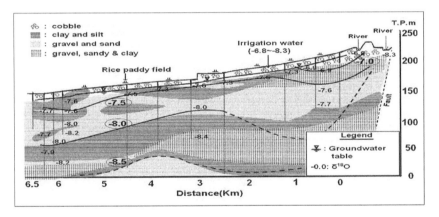

Figure 14.   Vertical distributions of $\delta^{18}O$ values (‰) of groundwater along the transect line B–B' on Left Bank. Thick line shows the isopleths of $\delta^{18}O$ value.

the rivers of Karyo-gawa and some tributaries and, for the left bank, by the Wanami-gawa or Echi-gawa.

6.3   *Characteristics of tritium and radon concentrations of groundwater*

Concentrations of tritium and radon in groundwater and surface water in this area vary widely. Tritium ranges from approximatly 1.0 to 6.6 T.U. with the highest frequency (exceeding 70% of all data) found in the range of 2.0~4.0 T.U. (Fig. 14). Radon exists between approximately 0 to 40 Beq/l. The residence time of these water samples was calculated by the piston flow model (Clarke and Fritz, 1997; Fig. 16). The tritium data for precipitation (4.1 ± 0.3 T.U.) was obtained from Tsukuba, Ibaraki Prefecture (Yabusaki et al., 2003).

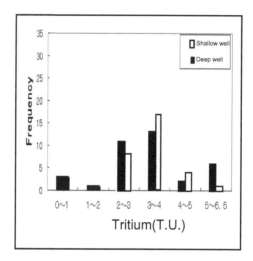

Figure 15.   Histogram of tritium concentrations of groundwater.

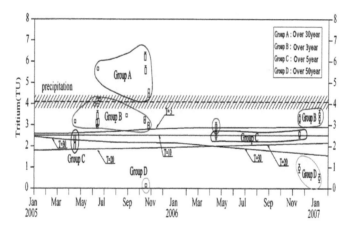

Figure 16.   Analytical results of tritium concentrations by the piston flow model.

6.4   *Vertical distribution of tritium and radon concentrations of groundwater*

As shown in the Fig. 16, the vertical change in the concentrations of tritium and radon are generally small. However, those concentrations generally decrease with increasing depth, i.e., the tritium concentration changes from approximately 3.0 T.U. to less than 1.0 T.U. and radon from approximately 25 Beq/l to nearly zero, being particularly very low in the deep layer. The residence time calculated from tritium concentrations of the groundwater in the aquifers are 3–5 years or longer in the shallow aquifer, approximately 30 years or longer in the medium aquifer and 50 years or longer in the deep-seated aquifer (water samples collected at depths of 80~90 m and 120 m), respectively.

These results indicate that the groundwater in the shallow and intermediate aquifers have a short cycle, i.e. water from the shallow aquifer permeates at a faster rate into the intermediate aquifer and mixes well, while that in the deep-seated aquifer flows at an

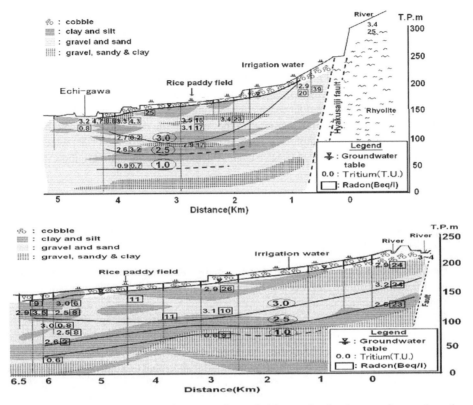

Figure 17. Vertical distributions of concentrations of tritium and radon in groundwater along the transect lines of A–A′ (upper) and B–B′ (lower). Thick lines show the isopleths of the tritium concentrations.

Table 1. Corrected ^{14}C ages of groundwater.

Sample No.	Depth (m bgl)	Aquifer	Corrected age (yr BP)
No. 37-1	−93.0~−95.3	Deep	3900 ± 30
No. 37-2	−31.0~−33.2	Medium	338 ± 27
No. 37-3	−11.5~−17.6	Shallow	Modern
H3-P	−92.0~−102.0	Deep	1865 ± 27
H3-(T-1)	−55.0~−78.0	Medium	283 ± 28
H3-(T-2)	−41.0~−44.2	Medium	43 ± 28
H3-(T-3)	−1.5 ~ −15.0	Shallow	708 ± 28

extremely low speed and is almost static. This result dose not conflict with the relationship between the concentration of SiO_2 and tritium, e.g., SiO_2 concentrations change according to the residence time (Haines and Lloyd, 1985), because the SiO_2 concentrations in the deep groundwater, particularly that stored in a clay layer, is generally higher in this region (at over 30 mg/l).

6.5   *Characteristics of ^{14}C concentrations of groundwater*

A total of seven samples of water were collected for determination of the ages of ground-water at different depths. The shallow groundwater is of recent age except for the shallow aquifer at a depth of 15 m (700y. B.P.; Table 1). Intermediate waters at depths ranging of from approximately 30–80 m originate about 100 years to 400 years ago, and deep-seated water at depths of more than 80 m is originated approximately 2,000–4,000 years ago. There are large differences in the residence time when compared to tritium, but the data revealed tendencies similar to the tritium analysis, so the water in the deep aquifer is old historical water. Note that the water sample at a depth of 15 m (sample No. H3-(T-3)) which shows an anomalous age, is considered to have been affected by intermediate groundwaters.

7   SUMMARY AND CONCLUSIONS

The summary of the findings of our survey on the groundwater flow system in the alluvial fan of the Echi-gawa is as follows.

1. Groundwater levels show seasonal changes, being highest in the rainy season and lowest in winter, and groundwater levels in the shallow aquifer are higher than those in the deep aquifer.
2. Shallow groundwater generally flows in a north-west direction, but flows toward the river channel around the apex of the fan and also separates into two directions near the residual hill in the downstream area.
3. The results of groundwater potentials in the aquifer show that groundwater in the shallow and intermediate aquifers moves vertically or downward on the right bank of the Echi-gawa, and mostly laterally on the left bank, but that in the deep-seated aquifer ground-water flows laterally on the right bank and moves laterally or upward on the left bank.
4. Water quality of most water samples is of a Ca-HCO$_3$ type except for the water at the piedmont area, in the deep aquifer (Na-HCO$_3$) and in spring water (Ca-SO$_4$). How-ever, the vertical changes in water quality are relatively small. The water quality of groundwater in the deep layer is similar to the water around the mountains.
5. The $\delta^{18}O$ values of groundwater range from −7.4 to −8.5‰. Low values are distributed around the mountain and high values (approximately −7.3‰) are distributed on the terrace plain at the centre of the fan. The $\delta^{18}O$ values of deep-seated groundwater are similar to those of the water around the mountain, as with the water quality.
6. Concentrations of stable and radioactive isotopes in the groundwater in the aquifer are generally similar; however, they decrease toward the deep layer, showing a significant change between the intermediate and deep-seated aquifers (depth exceeding 80 m). In particular, concentrations of tritium and radon in the deep aquifer are extremely low.
7. Tritium concentrations of groundwater in the aquifer are generally divided into three groups, from the shallow aquifer toward the deep-seated aquifer, i.e., approximately 3.0 T.U., 3.0–1.0 T.U. and less than 1.0 T.U., respectively. The residence times of the groups are 3–5 years in the shallow aquifer, 30 years or longer in the intermediate aquifer and 50 years or longer in the deep-seated aquifer.

These results indicate that the shallow groundwater originated from precipitation and irrigation water penetrating into the deeper layers except in the area at the centre of the fan on the Left Bank, and that groundwater in the shallow and intermediate aquifers actively circulate and mix with water flowing vertically or laterally in the groundwater system. On the other hand, groundwater in the deep aquifer is recharged by groundwater around the intermountain area and flows in a downstream direction at an extremely low speed and is almost static.

ACKNOWLEDGEMENTS

We extend our deepest gratitude to the staff of the Higashi-omi City Water Service Office and the Echi County Water Service Office for their great support and cooperation in the survey activities and their permission to disclose data. Moreover, we wish to express our gratitude to the students of Osaka Kyoiku University, Laboratory of Hydrological Sciences, for their considerable cooperation with the field survey activities and for their assistance with the data reduction.

REFERENCES

Clarke, I.D., Fritz, P. (1997) Environmental Isotopes in Hydrology. Lewis Publisher, New York, 171–196.

Foundation of Nature Conservation in Shiga Prefecture (1979) Land and Life in Shiga, 44–49 (in Japan with English abstract).

Haines, T.S., Lloyd, J.W. (1985) Controls on silica in groundwater environments in the United Kingdom. Jour., 81, 277–295.

Higashi-Omi City (2007) Report of groundwater survey, 2007.

Horino, H., Maruyama, T. (1990) Spatial fluctuation characteristic of fluctuation band of an annual groundwater table, Jour. Groundwater, 32(2), 81–90 (in Japan).

Horino, H., Watanabe, S., Maruyama, T. (1989) Study of an Actual Demonstrations on the Role of Groundwater in Agricultural Water Utilization. Paper series of agricultural engineering, 144, 9–16 (in Japan).

Horiuchi, K., Murakami, Y. (1977) Basic studies on the new determination method of radon in mineral springs with a liquid scintillation counter, J. Balneo. Soc. Japan, 28, 39–47.

Kobayashi, M., Hamada, T., Takahata, M. (2007) Characteristic of Water Quality and Runoff Mechanism of Spring Water in the alluvial fan of riv. Echi-gawa. Jour. Groundwater Technology, 49(3), 9 (in Japan).

Motokage, I., Minami, I. (1975) A Numerical Method for Simulating regional Groundwater Level at Riv. Echi Delta zone, Trans. JSIDRE, 58, 53–77 (in Japan).

Oishi, A., Kobayashi, M., Hamada, T., Okuda, E., Miyazaki, S., Hu, Sung Gi (2008) Hydrogeology on Echi-gawa Alluvial Fan, Shiga Prefecture, Research group on Hydro- environment around alluvial Fans, 191–210.

Takaya, Y., Nishida, K. (1964) Geology and ground water discharge mechanism at east side of Biwako, Southwest Japan, Earth Science (Chikyu Kagaku), 74, 33–39. (in Japan with English abstract).

Uemura, Y. (1979) The paleogeomorphology and historical development of landforms of the Koto hills, Ritsumeikan Bungaku, 410/411, 143–174 (in Japan).

Yabusaki, S., Tsujimura, M., Tase, N. (2003) Seasonal Change of Tritium Concentrations of precipitation in the Kanto District. Bulletin of the Terrestrial Environments Research Center the Tsukuba University, 4, 11, 9–124 (in Japan).

CHAPTER 15

Modelling of groundwater flow characterized by hydrogeological structures and interaction with surface water at Shigenobu-gawa alluvial fan, Ehime Prefecture, Japan, with a preliminary examination of the influence of climate change

Osamu Watanabe
Watanabe Suimon Kikaku, Katsura, Izumi-ku, Sendai, Japan

Tomoaki Kayaki
K-HGS Co., Ltd., Hatsukaichi, Hatsukaichi-shi, Hiroshima, Japan

Hiroji Ichimaru
I-Data, Kuramatsu, Sugito-machi, Kitakatsusika-gun, Saitama, Japan

Takaaki Hijii
Geo Field Survey, Maidashi, Higashi-ku, Fukuoka, Japan

ABSTRACT: Shallow groundwater is the primary water source of the Dogo Plain, and it depends on infiltration from the Shigenobu-gawa river. This research attempts to model this groundwater system, paying attention to the distribution of the *segire* (dry riverbed areas) and *izumi* (springs) along the Shigenobu-gawa. Although it was a preliminary examination based on limited data, it has resulted in a model being built that can reproduce the *segire* and *izumi* distribution, reflecting the local hydrogeological structure. A preliminary examination of the influence of climate change using this model was also carried out. A fall in the groundwater table accompanying changes in recharge conditions was predicted. This suggests that, to prepare for the future, it is necessary to clarify the details of the hydrogeological features of the deep aquifer of the Dogo Plain.

Keywords: Dogo Plain, groundwater—surface water interaction, buried tectonic basin, groundwater modelling, climate change

1 INTRODUCTION

The Dogo Plain is situated at the western end of Shikoku Island, occupying an area of nearly 73 km^2. Currently, 600,000 people live on this plain. A large amount of groundwater is used to supply cities, industry and agriculture on the Dogo Plain. Matsuyama City depends on groundwater for approximately 52% of its tap water, and another two cities and two towns depend on groundwater for almost all of their tap water (Masaki-cho et al., 2007; Matsuyama City, 2007; Toon City, 2007).

The Shigenobu-gawa river, which flows through this plain, is 36 km long with a catchment area of 445 km^2. According to an existing report (Hida, 1978) about the water budget of the Shigenobu-gawa valley, the volume of infiltrated water from the downstream section of the Shigenobu-gawa river is almost equal to the abstraction capability of the public-water and industrial-water wells in the downstream area. Moreover, the Shigenobu-gawa is known for "*segire*," which are normal occurrences of riverbed exposure occurring due to lack of surface water in ephemeral channels. Spring water, called "*izumi*," is also distributed along the Shigenobu-gawa, and has been used for paddy field irrigation and other uses for many years. As such, most of the water resources in the Dogo Plain depend on the shallow groundwater which is recharged by the Shigenobu-gawa.

According to the RCM20 (Regional Climate Model with a horizontal resolution of 20 km) predictions to 2100 for the A2 emissions scenario (Japan Meteorological Agency, 2005), the monthly precipitation in the drought season will decrease at the Dogo Plain. This change in precipitation pattern will have a significant influence on the flow regime of the tributaries, leading to a decrease of groundwater recharge via the river bed of the Shigenobu-gawa. Since most of the water resources here depend on shallow groundwater supplied from the river, it is feared that the predicted change in the flow regime will cause an acute disruption to the local water supply.

In this research, we first investigated the current inter-relationship of the river and ground-water, the groundwater recharge mechanisms, and groundwater flow systems. Then, we used simulations based on groundwater modelling to conduct a preliminary examination of the influence of precipitation change due to global warming.

2   OUTLINE OF THE STUDY AREA

2.1   *Geomorphology and hydrogeology of the Dogo Plain*

The Shigenobu-gawa river, originating from Higashi Mikataga-hara (1,233 m in height) in the Takanawa mountain range of western Shikoku Island, flows southward until it meets the Omote-gawa river and then turns east and flows into Iyo-nada, after joining the Tobe-gawa river flowing from the southwest and the Ishite-gawa from the northeast. The area from the point of confluence with the Omote-gawa to the sea is called the Dogo Plain, Matsuyama Plain, or Shigenobu-gawa Fan due to its fan-like topography. Fig. 1 shows the location and topographical classification of the study area.

The Cretaceous Izumi Group is widely distributed and forms the basement for much of the Shigenobu-gawa river catchment area. The later deposits distributed on the Shigenobu lowlands are gravel beds with silt lens layers with a thickness of ≥600 m. According to the Geotechnical Map of the Matsuyama Plain (Ehime Construction Research Institute, 2003), surface layers up to a depth of around 30 m below ground level (m bgl) consist of alluvium, while those below 30 m bgl are pre-alluvium (Kashima & Takahashi, 1980). The alluvium consists of sandstone gravels from the Izumi Group with volcanic rock pebbles and granite pebbles (Table 1). The surface layers are subrounded—rounded gravel with a poor matrix, but are rich in voids, particularly near the current and former riverbeds. Discontinuous silt lenses concealed 20–30 m bgl form the lowermost layer of the alluvium (Miyazaki et al., 2008). The pre-alluvium gravel beds are poorly characterized, but it is likely that they are correlated with the old alluvial fan deposits.

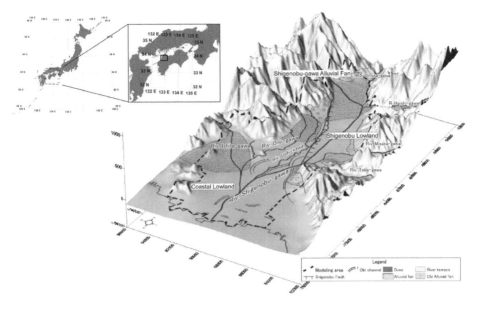

Figure 1.   Location and topographical classification of the study area.

Table 1. Hydrogeological correlation table of the Dogo Plain (Miyazaki et al., 2008).

Geologic time			Formation	Stratigraphic Facies			Aquifer	Depth
				Coastal Lowland	Shigenobu Lowland	Shigenobu-gawa Alluvial Fan and Others		
Cenozoic	Quaternary	Holocene	Alluvium	Sand bed (intercalated silt layer)	Rounded-gravel bed with many voids (2.0x10⁻³ to 4.1x10⁻² m/sec)	Fluvial Sediment (river bed)	Shallow	
				Gravel bed (intercalated silt layer)	Silt layer (lens-like thin layer)			20-30m
		Pleistocene	Old Alluvial Fan	Debris flow deposit (Fluvial sediment, Floodplain deposit)			Deep	
				(1.0x10⁻⁵m/sec -)	(1.0x10⁻⁴m/sec -)			
	Neogene	Pliocene	Second-Setouchi Group	Fluvial sediment (poor in the fine-grained matrix)	-			600m
			(Ishizuchi Group)	—			—	
	Paleogene	Eocene	(Kuma Group)	—				
Mesozoic/Cretaceous			Izumi Group	Sandstones and Mudstone			Impermeable Basement	

Fig. 2 shows contour maps of the base of the shallow alluvium and the upper surface of the Dogo Plain basement, which were elucidated using existing test wells and geophysical prospecting (Goto et al., 1999; Ikeda et al., 2003). Fig. 3 shows the longitudinal profile of the base of the alluvium and the upper surface of the basement along the Shigenobu-gawa.

The Shigenobu lowland is a buried tectonic basin, which buried the north-slip half graben formed by the Shigenobu fault. The buried tectonic basin between the Omote-gawa and the Tobe-gawa river has a length of approximately 8 km (east–west) and a width of approximately 4 km (north–south). The thickness of the deposit is 600 m (Fig. 2).

In the coastal lowland, in the area extending downstream from the Tobe-gawa river, the thickness of the gravel bed decreases toward the coast. The characteristics of the shallow

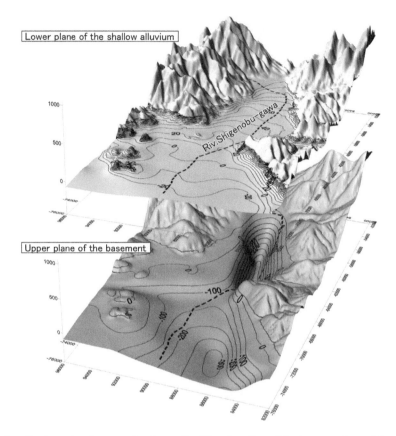

Figure 2.   Contours (m above sea level, m asl) of the elevation of the lower surface of the shallow alluvium (upper panel) and upper surface of the basement (lower panel).

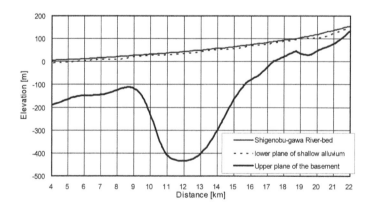

Figure 3.   Longitudinal profile of the elevation (m asl) of the base of the alluvium and the upper surface of the basement along the Shigenobu-gawa.

aquifers differ between the Shigenobu and the coastal lowland areas, with the distribution of wells in the coastal lowland becoming increasingly sparse (Miyazaki et al., 2008).

2.2 *Water budget in the Dogo Plain*

Groundwater in the Dogo Plain is abstracted from 79 wells for city and industrial use. These pumping wells mainly use shallow groundwater at 10–15 m depth, except for some wells downstream. The groundwater abstraction rate of these wells is approximately 47,000,000 m³/year in Matsuyama City and 11,000,000 m³/year in the two neighbouring cities and one town. However, these represent only the abstractions controlled by the local governing body for city and industrial use, and there appears to be additional groundwater withdrawal by factories and for agricultural irrigation.

The middle panel of Fig. 4 compares groundwater withdrawal rate (Masaki-cho, 2007; Matsuyama City, 2007; Toon City, 2007) with a hydrograph of the total discharge rate of

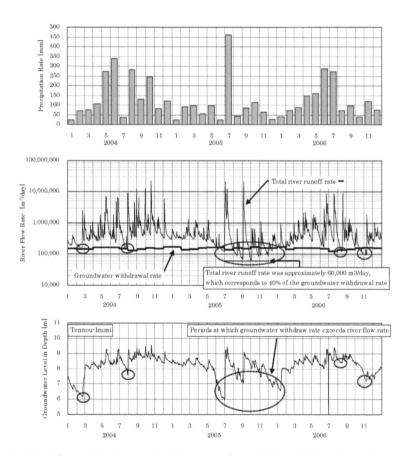

Figure 4.   Monthly precipitation rate in Matsuyama from 2004 to 2006 (top), total river runoff rate into the Dogo Plain (from on a runoff-analysis using the tank model method based on actual measurements quoted from the Water Information System) and groundwater withdrawal rate (middle), and groundwater water level at Tennou-Izumi (bottom).

the four rivers calculated by runoff analysis. The tank model of the typical Shigenobu-gawa, Omote-gawa, Tobe-gawa, and Misaka-gawa flow into the Dogo Plain was created based on the actual measurements quoted from the Water Information System (Ministry of Land Infrastructure and Transport Japan, 2008). Groundwater is always withdrawn at approximately 150,000 m^3/day. However, the total base flow rate of the four main rivers decreases when the precipitation rate is low. In particular, in September and October 2005, the total river runoff rate was approximately 60,000 m^3/day, which corresponds to 40% of the groundwater withdrawal rate of 150,000 m^3/day.

The bottom of Fig. 4 shows the changes in groundwater level of the Tenno-Izumi well in the downstream section. The change in groundwater level follows the hydrograph, showing that rapidly circulating groundwater is flowing in high-permeability alluvial deposits. However, the groundwater level shows a rapid decline from January to February 2004, from June to December 2005, and from November to December 2006, so it is not completely synchronized with the hydrograph. During these periods, the groundwater withdrawal rate is larger than the river flow rate. Considering these conditions, the groundwater in the Dogo Plain is recharged by the river, and it shows the direct influence of river flow changes.

3   METHODOLOGY

3.1   *Method for studying the groundwater flow system*

To understand groundwater flow in the Dogo Plain, simultaneous observation of the groundwater table elevations was carried out during October 23–24, 2007. Since this period was a drought and non-irrigation period, it was comparatively easy to see the groundwater flow situation. Existing wells and *izumi*, which are distributed within the Shigenobu-gawa, were selected for observation. Moreover, water quality analysis was conducted at representative points.

In addition, simultaneous observation of the river flow rate aimed at elucidating the groundwater recharge mechanism from Shigenobu-gawa was carried out over the same period. We measured the Shigenobu-gawa river flow at intervals of about 1 km, and examined the sectional river channel water budget using this result.

3.2   *Groundwater modelling*

3.2.1   *Model geometry*
An area of 147.15 km^2 of the Dogo Plain, including alluvial fan areas such as those of the rivers Shigenobu-gawa, Ishite-gawa, Misaka-gawa, and Tobe-gawa was selected for our investigation (Fig. 5), and a groundwater flow model created for this area. The FEFLOW software (WASY Software) was used to conduct a steady-state, saturated groundwater flow analysis in a three-dimensional finite element method mesh. To create the mesh, the horizontal zone of investigation was split into 11,288 triangulated elements based on considerations of the main river networks and their geographical zone boundaries, and the distribution of springs and wells in this area. The main aquifers of the Dogo Plain can be subdivided into alluvium deposits and the underlying old-alluvial fan deposits. These deposits were split into two and three model layers, respectively, to create a five-layer structure that made it possible to set the vertical hydraulic parameter distribution.

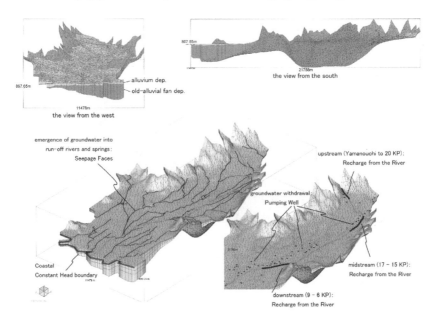

Figure 5.   Groundwater flow model (aquifer, groundwater basin shape, and boundary conditions).

3.2.2   *Boundary conditions of the groundwater model*

Precipitation-based recharge and recharge by leakage through the riverbed of Shigenobu-gawa were set as the groundwater recharge boundary conditions. For the latter, recharge locations were taken, on the basis of current conditions, as being the following three areas: upstream, midstream, and downstream. Recharge amounts for each area were set as the volumes of infiltrated water, calculated from the results of simultaneous river flow rate observations carried out on October 24–25, 2007.

In addition, we set coastal constant head boundaries and seepage faces to simulate the emergence of groundwater into run-off rivers and springs. We also set the usage of groundwater from water resource wells and springs as pumping wells (Fig. 5).

3.2.3   *Hydraulic parameters of the groundwater model*

We set initial values for the hydraulic parameters as shown in Fig. 6 for each range classified hydrogeologically or geographically. The hydraulic conductivities of the shallow alluvium deposits were set as being of the order of 10^{-3} m/s (Shigenobu-gawa alluvial fan: 1×10^{-3} m/s, Shigenobu-gawa midstream to downstream: 3×10^{-3} m/s, etc.). Areas in which abstractions are concentrated and areas in which former riverbeds are located were given a higher hydraulic conductivity (on the order of 10^{-2} m/s) on the basis of existing geological survey results for water resource well locations.

On the other hand, only few investigations about old alluvial fan deposits, such as deep boring for hydrogeological survey, have been made. According to the limited survey results available, a hydraulic conductivity of the order of 10^{-5} m/s may be appropriate for the old-alluvial fan deposit (debris flow deposit). In our model, the hydraulic conductivity for the upper layer of the old-alluvial fan deposit was set as the order of 10^{-4} m/s and the value for the lower layer of the same deposit was set as the order of 10^{-5} m/s.

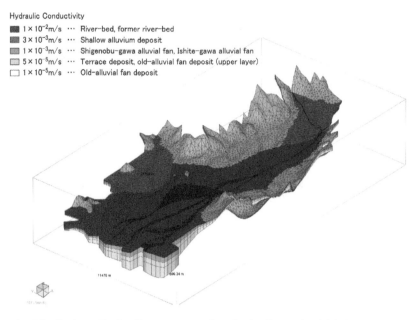

Figure 6.   Distribution of hydraulic parameter values (hydraulic conductivities).

Terrace deposits and old alluvial fan deposits distributed toward the edges of the plain were judged to have relatively poor permeability on the basis of their date of formation and the quality of groundwater that flows through them. Their hydraulic conductivities were therefore set as being of the order of 10^{-5} m/s, the same as for deep old-alluvial fan deposits.

3.3   Examination of the influence of climate change

3.3.1   Future prospects of climate change around the Shigenobu-gawa alluvial fan

Results from the RCM20 based on the SRES A2 scenario (Japan Meteorological Agency, 2005) were used as future meteorological conditions. According to the RCM20 predictions to 2100, average monthly precipitation in summer is forecast to increase, whereas average monthly precipitation in winter is expected to decrease. Table 2 shows changes in recharge conditions predicted for the winter period from October to March. Average winter (Oct–Mar) precipitation will fall to approximately 80% of its current value over the next 100 years. Moreover, considering the increase in potential evapotranspiration caused by the higher winter temperatures, it can be seen that the recharge potential amount (P-ET) is forecast to fall to 52% of its current value.

3.3.2   Numerical study of the change in the hydrological environment in the Shigenobu-gawa alluvial fan caused by climate change

The flow regime of the rivers that flow into the Dogo Plain is predicted to change with reductions in precipitation. In this study, runoff analysis using the tank model method for the main rivers was conducted, and the change in flow regime accompanying changes in

Table 2. RCM20 climate change forecasts (data for October to March only).

	RCM20 predictions		
	1981–2000	2031–2050	2081–2100
Mean temperature (Oct–Mar) [°C]	10.3	12.9	13.2
Potential evapo-transpiration (Oct–Mar) : ET [mm/y]	350.1	417.9	431.3
Mean rainfall intensity (Oct–Mar) : P [mm/y]	884.2	863.0	710.2
P-ET [mm/y]	534.1	445.2	278.9
change rate of "P-ET"		83.3%	52.2%

precipitation was considered. Then, by a simulation that set up the amount of groundwater recharge assumed from the runoff analysis result, the change in the groundwater table elevation caused by climate change was investigated.

4   RESULTS

4.1   *Groundwater systems based on the distribution of groundwater table elevation*

Fig. 7 shows the groundwater table elevation based on the results of simultaneous groundwater observation. The distribution of *segire* (dry riverbed) sections and the position of abstraction wells are also shown. The groundwater flow situation around Shigenobu-gawa shown in this figure is described below.

- In this period, the *segire* sections were distributed in three places: (1) the upstream section located in the centre of the Shigenobu-gawa alluvial fan, (2) the midstream section from 16 KP (the lower stream of the Omote-gawa juncture) to 10 KP (the upper stream of the Tobe-gawa juncture), and (3) the section downstream of the Tobe-gawa juncture.
- Near the upper reaches of these *segire* sections, the groundwater table was low, compared with river level, showing that the groundwater had been recharged from the river.
- From the distribution of the groundwater table elevation, the existence of three groundwater flow systems shown in the figure is indicated.

Fig. 8 shows the longitudinal profile of the groundwater table and the Shigenobu-gawa riverbed at the time of the simultaneous observations. The hydrogeological structure in the Shigenobu-gawa is also shown. The longitudinal profile of the groundwater table shows that the curve follows that of the riverbed from the mouth of the river to the Tobe-gawa juncture near 10 KP. However, in the section around 15 KP, the longitudinal profile of the groundwater table separates from that of the bed slope, and the groundwater table is lower than the riverbed by about 10 m. This section corresponds with the *segire* midstream section (from 16 to 10 KP). It is thought that this *segire* section of the Shigenobu-gawa is caused by a large-scale buried structural basin, which was formed by the Shigenobu Fault and related structures. Moreover, it is thought that the distribution of *izumi* concentrated near the Tobe-gawa juncture was caused by a groundwater flow that has been dammed by the rise of the Izumi Group basement (geologic structure barrier) near the Tobe-gawa juncture.

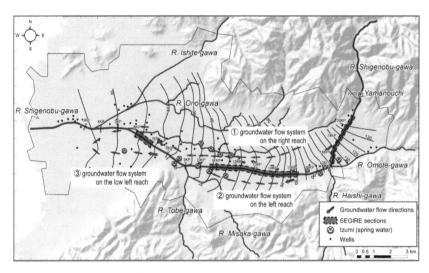

Figure 7.　Equipotential map of groundwater table elevation (m asl) and distribution of *segire* sections and *izumi* during the observation (October 23–24, 2007).

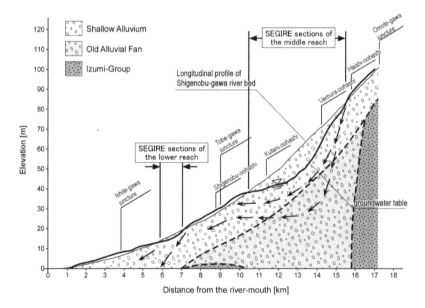

Figure 8.　Longitudinal profile of Shigenobu-gawa riverbed and groundwater table (m asl).

4.2　*Sectional river channel water budget*

To study the actual underground flow rate for the recharging of groundwater from Shigenobu-gawa, we examined the sectional river channel water budget, as shown in Fig. 9. The results showed the following:

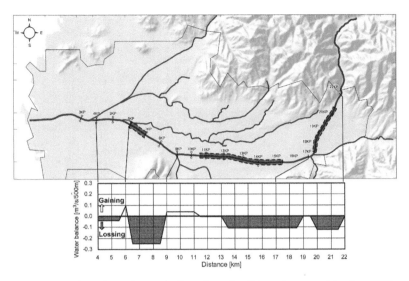

Figure 9.   Sectional river channel water budget along the Shigenobu-gawa (October 2007).

- The flow rate of Shigenobu-gawa was 0.8 m³/s at 22 KP, where the Shigenobu-gawa flows out from the mountainous area into the plain. Most of this river flow was lost near 21KP, and the *segire* has appeared in the section from there to 17KP.
- The flow rate after the Omote-gawa juncture was 0.6 m³/s, but there was no surface flow at approximately 1.5 km after the juncture (approximately 15.5 KP). Including the inflow from the other river branches in this 1.5 km reach, the underground flow in the section from the Omote-gawa juncture to the midstream *segire* section has a flow rate of 1.2 m³/s.
- The Tobe-gawa joins the Shigenobu-gawa at approximately 9 KP, so that the flow rate of the Shigenobu-gawa was 0.4 m³/s at 8 KP. After that, the river flow decreased continuously and finally reached zero at 6.5 KP, which was the downstream *segire* section.
- The flow rate at 3 KP after the Ishite-gawa juncture was 1.3 m³/s due to inflows from the Ono-gawa and the drainage channels.

It can be understood that the change in river flow in each of these sections accompanies the distribution of groundwater table elevation resulting from the hydrogeological structure along the Shigenobu-gawa shown in Fig. 8.

4.3   *Groundwater flow systems based on the results of water quality analysis*

Fig. 10 shows the distribution of groundwater quality (as hexagonal diagrams) for representative points, and the interpolated distribution of electrical conductivity.

The electrical conductivity of the water of the main feeder rivers, such as the Shigenobu-gawa and Omote-gawa, was approximately 250 μS/cm. Except for the Ono-gawa valley and the area around the mouth of the river, the electrical conductivity of the groundwater and spring water was 200–300 μS/cm. Thus, it was almost the same as that of the river water. This electrical conductivity distribution also shows the close relationships between groundwater flow and the river water.

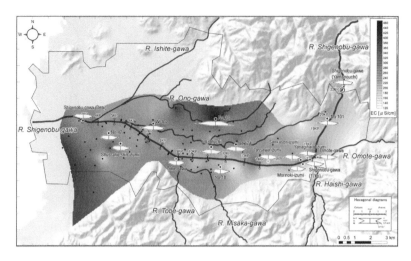

Figure 10.   Distribution of groundwater quality (shown as Hexagonal diagrams for selected wells) and Electrical Conductivity in the Dogo Plain.

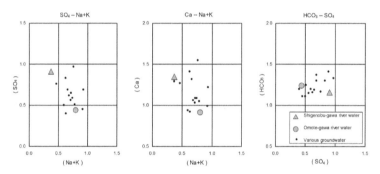

Figure 11.   Scatter plot showing correlation between Na + K and SO$_4$, Na + K and Ca, and HCO$_3$ and SO$_4$ for various groundwater samples.

The river water and groundwater in the Dogo Plain are alkaline with similar levels of Ca(HCO$_3$)$_2$. It is supposed that this is distributed via quickly circulating groundwater. The water of the Shigenobu-gawa and Omote-gawa also have similar levels of Ca(HCO$_3$)$_2$, but that of Shigenobu-gawa has extremely high levels of Na + K and a lower level of SO$_4$ than that of the Omote-gawa. Fig. 11 shows the relationship between Na + K and SO$_4$, Na + K and Ca, and HCO$_3$ and SO$_4$ in the river water of the Shigenobu-gawa (indicated by gray triangles) and Omote-gawa (gray circles). According to this figure, the chemical composition of other groundwater and spring water samples are generally between those of the Shigenobu-gawa and Omote-gawa river waters; therefore, the shallow groundwater in the Dogo Plain may consist of groundwater mixed with the two types of river water.

Fig. 12 shows the groundwater flow system in the Dogo Plain based on the results of the water quality analyses. The water quality of the groundwater in the upstream area of the Shigenobu-gawa is almost the same as the Shigenobu-gawa river water. However, the groundwater downstream from the Omote-gawa juncture contains about 60 to 70%

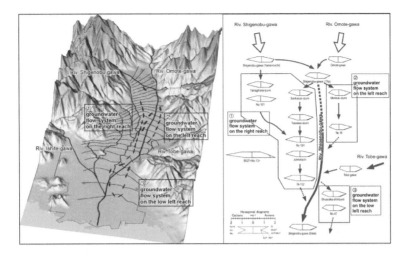

Figure 12. Groundwater flow system based on the distribution of groundwater table elevation and the result of water quality analysis.

Omote-gawa river water. These tendencies are in agreement with the flow systems based on groundwater table elevation distribution.

4.4 *Groundwater modelling (calibration and verification of groundwater model)*

The groundwater model was calibrated against the groundwater table observations conducted on October 24–25, 2007 by adjusting the hydraulic conductivities. In this study, a steady-state analysis under the average recharge conditions (Table 3) was carried out. Fig. 13 shows that the calibrated model can mostly reproduce the global shape of the observed groundwater table. In this model, by increasing the hydraulic conductivity of the upper layer of the old-alluvial fan deposits in the midstream section of Shigenobu-gawa, we increased the aquifer thickness in the buried structural basin, which is thought to contain a greater concentration of debris flow deposits than the surrounding area (Fig. 14). As a result, we can see from the *segire* along the midstream of the Shigenobu-gawa and the distribution of spring outlets seen in the vicinity of the juncture with the Tobe-gawa that the observed shape of the groundwater table in the area has been well reproduced by the model.

Fig. 15 displays the difference in the groundwater table elevations between the initial and calibrated models. In the area bounded by the dashed line on the figure, the buried structure is at its deepest, and corresponds to the location given a high hydraulic conductivity value for the old-alluvial fan deposit in the calibrated model. It is clear that, on the upstream side of this area, the groundwater level is lower than that obtained by the initial model, and that the groundwater level downstream of this area is higher. This place, where the rise of the water table was reproduced with the calibrated model, agrees with the region over which a considerable amount of spring water is distributed. The photograph in the figure shows the Jo-no-Fuchi Park as a representative case of spring water located there. By taking hydrogeological structure into consideration, it became possible to reproduce a *segire* and an *izumi*.

Table 3.   Model verification results against observed groundwater levels.

Case	Meteorological condition (P-ET) [mm/day]	Correlation coefficient	RMSE [m]	Number of observation points
Oct-2007	1.09	0.995	2.079	96
Jan-2004	0.66	0.996	2.042	46
Feb-2004	0.12	0.993	2.419	41
Mar-2004	1.88	0.998	1.834	41

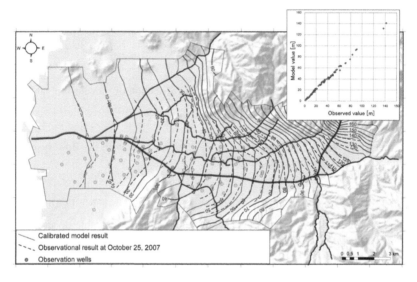

Figure 13.   Contour map (m asl) for the current groundwater table and simulation results of the calibrated model.

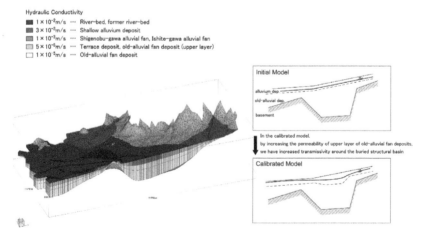

Figure 14.   Hydraulic conductivity values used in the calibrated model.

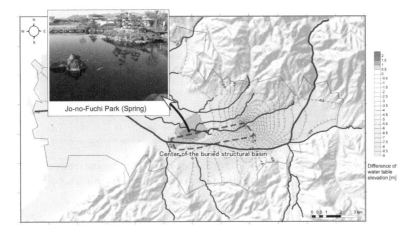

Figure 15.   Difference between the groundwater tables obtained with the initial and calibrated models (Appearance of *segire* and *izumi* that were realized by consideration of hydrogeologic structure).

Moreover, the model was verified by checking the model results for three cases (January, February, and March 2004) where recharge conditions differ. A model validation result against observed groundwater levels is shown in Table 3. It is considered that the model has been verified across a sufficient a range of recharge conditions (considering the average recharge of 1.46 mm/day in 1981–2000, and 0.76 mm/day in 2081–2100).

5   DISCUSSION

5.1   *Preliminary examination of the influence of climate change*

We examined flow rates under the changes predicted to occur as a result of climate change using a flow model (tank model) created for the main rivers feeding into the plain in the area under investigation. Fig. 16a shows the discharge duration curves that result when the RCM20 predictions of precipitation and temperature for the mountainous region in the vicinity of the catchment areas of the tributaries (Yamanouchi region) between October and March are used as input for the tank model created for the upstream region of Shigenobu-gawa. This result clearly shows the decreased flow rate predicted for that time.

Along with the decrease in the flow rates of the tributaries, a decrease is predicted in the infiltration flow through the riverbed of the Shigenobu-gawa (the primary groundwater recharge mechanism in the Dogo Plain). Fig. 16b shows the relationship between the upstream flow rate (at Yamanouchi) and the infiltration amount for the upstream section of the Shigenobu-gawa. Using this approximate formula, it is possible to empirically predict the changes in the infiltration flow at the upstream section of the Shigenobu-gawa based on predicted changes in the upstream flow rate. By applying the same method to the other tributaries, the following changes in recharge conditions can be forecast:

- upstream section: The average flow rate of the Shigenobu-gawa in the period from October to March will fall to 85% of its current value, and the upstream infiltration flow to 61% of its current value.

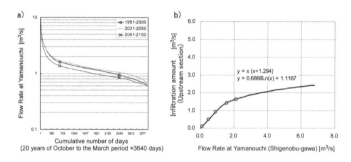

Figure 16.   a) Yamanouchi flow regime (October to March) under current and future climate based on SRES A2 scenario, b) Relationship between the Yamanouchi flow and the Shigenobu-gawa upstream section infiltration flow.

- midstream section: The average flow rate of the Omote-gawa in the period from October to March will fall to 77% of its current value, and the midstream infiltration flow to 84% of its current value.
- downstream section: The average flow rate of the Tobe-gawa and Misaka-gawa in the period from October to March will fall to 82% of its current value, and the downstream infiltration flow to 94% of its current value.

Using the groundwater model, we conducted a simulation with the above recharge conditions. In this manner, we investigated the changes in groundwater table elevation expected to result from climate change. According to the result shown in Fig. 17, it was predicted that a remarkable groundwater table elevation fall occurs around the head of the Shigenobu-gawa alluvial fan (2.0–2.5 m) and the *segire* regions (0.5–1.0 m). This fall will be due to a reduction in precipitation during the drought period (October to March), an increase in evapotranspiration resulting from global warming, and attendant reductions in the flow rates of major tributaries.

5.2   *Possible countermeasures and future studies*

Most of the groundwater used in the Dogo Plain is from springs and shallow wells. Accordingly, there is concern that the fall in shallow groundwater table elevation that will accompany worsening recharge conditions will affect water availability in the region. Therefore, it is crucial to adopt methods for judicious use of fresh water resources.

Although both the Omogo and Ishide dams were constructed to provide fresh water resources other than river water and groundwater in the Dogo Plain, their limited storage capacity restricts their use to an optimum level during the high-demand seasons. Moreover, these reservoirs are also directly affected by climate change. Therefore, for the sustainable use of both surface water and groundwater, it is necessary to take measures to withstand climatic changes. For example, enhancing the retention capacity of springs or artificial ponds and extending paddy fields and preventing their further reduction are some possible measures. Developing new water retention areas and using fresh water resources effectively are urgent requirements.

According to Miyazaki et al. (2008), a large-scale collapsed water basin consisting of two-storied aquifers can be observed in the middle part of the river (Fig. 2). The upper aquifer, formed in the Holocene period, stores nearly 110 million m³ of groundwater, while

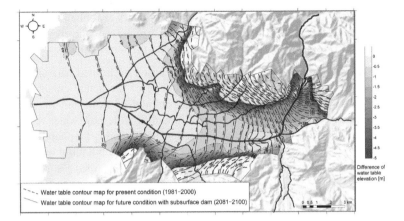

Figure 17.   Simulated water table elevation based on recharge conditions for 2000 and 2100 from RCM20 simualtion.

the lower Pleistocene aquifer possibly stores 1,500 million m^3 of groundwater, assuming 0.1% effective porosity, although this figure is uncertain. If these aquifers can be developed, a majority of the future water problems in the Dogo Plain will be completely solved. Therefore, to address future water problems in this region, it is necessary to clarify the exact nature, size, capacity, and water quality of the entire sedimentary basin.

6   SUMMARY

In this research, we investigated the current interrelationship between the river and groundwater, the groundwater recharge mechanisms, and groundwater flow systems on the basis of simultaneous groundwater table elevations and river flow rate observations. In addition, a groundwater flow mechanism based on the hydrogeological structure was considered. Next, we investigated the changes that can be expected in the hydrologic environment in future by constructing a groundwater model that reflects the hydrogeological structures and the mechanism of interaction with surface water. We summarize our results below.

- The shallow groundwater in the Dogo Plain consists of quickly circulating subsurface water that infiltrates from the upstream of the *segire* sections of the Shigenobu-gawa river. The subsurface water flow can be classified into three large groups according to their flowpaths.
- The appearance of a *segire* in Shigenobu-gawa is an example of groundwater and surface water interactions. It can be understood that the reduction in the river flow in each of these sections arises from changes in the groundwater table elevation resulting from the hydrogeological structure along the Shigenobu-gawa.
- We have created a groundwater flow model that can recreate current conditions. With this model, we can also recreate the shape of the groundwater table in the area by setting hydraulic conductivity parameters so as to reflect the shape of the buried structural basin in the midstream area of the Shigenobu-gawa. The accuracy of the results is demonstrated by the distribution of spring outlets seen in the vicinity of the Tobe-gawa junction.

- From the result of the simulation carried out as a preliminary examination of the influence of climate change, we predict that a dramatic fall in groundwater levels will occur at the head of the Shigenobu-gawa alluvial fan and around the *segire* regions.

The groundwater model used in this research is based on limited data and the results of a recent investigation. A more clear understanding of the actual conditions of deep groundwater flow, such as groundwater storage, amount of flow, and water quality, is necessary to clearly show whether a deep well could provide new water resources. We think that a more accurate report could be attained by investigating the shape (depth) of the groundwater basin and the hydrogeological features of the deep aquifer and by carrying out examinations using the groundwater model based on the obtained results.

ACKNOWLEDGMENTS

The authors wish to express their sincere thanks to the municipalities of Matsuyama city, Masaki town, Toon city, and others in the Dogo Plain for their primary study of data related to water and groundwater issues in the region. The authors also thank Mr. Koike, Manager of the River Division of the Shikoku branch of MLIT, for his kind advice on this work. The authors are grateful to the committee members and colleagues of RHF for their continuous encouragement and helpful advice on this work.

REFERENCES

Ehime Construction Research Institute (2003) Geotechnical Map of the Matsuyama Plain, 2003.

Goto, H., Nakata, T., Okumura, K., Ikeuchi, A., Kuramura, Y., Takada, K. (1999) Holocene Faulting of the Shigenobu Fault, the Median Tectonic Line Active Fault System, West Shikoku, Japan. *Geography Rev. Japan Ser. A*, 72, 267–279.

Hida, N. (1978) Groundwater use at Shigenobu-gawa downstream area, Ehime Pref., Water budget in Japan, 230–241.

Ikeda, M., Ohno, I., Ohno, Y., Okada, A. (2003) Subsurface Structure and Fault Segmentation along the Median Tectonic Line Active Fault System, Northwestern Shikoku, Japan. *J. Seismo. Soc. Japan*, 56, 141–155.

Japan Meteorological Agency (2005) Global Warming Projection Vol. 6, http://ds.data.jma.go.jp/tcc/tcc/products/gwp/gwp6/index.htm

Kashima, N., Takahashi, J. (1980) Environmental-geologic Study of the Matsuyama Plain, Shikoku—1. Regional Geology of the Matsuyama Plain. *Mem. Ehime Univ. Nat. Sci. Ser.D (Earth Science)*, 9, 3, 63–72.

Masaki-cho, Iyo-gun, Ehime Pref. (2007) Groundwater Observation Data, Drinking Water Source Quantity-of-water-intake Data.

Matsuyama City, Ehime Pref. (2007) Data about a Shigenobu-gawa Drainage System Permission Water Source.

Ministry of Land Infrastructure and Transport Japan (2008) Water Information System, http://www1.river.go.jp/.

Miyazaki S., Hasegawa S., Kayaki T., Watanabe O. (2008) Hydrogeology of the Shigenobu-gawa Alluvial Fan, Ehime Pref., Shikoku, Japan., "Hydro-environments of Alluvial Fans in Japan" Monograph, 36th IAH Cong. 2008 Toyama.

Toon City, Ehime Pref. (2007) Potable-water Water Resources Data.

CHAPTER 16

Hydrogeology and water balance in R. Chikugo-gawa Plain, Fukuoka Prefecture, Japan

Satoshi Hasegawa, Akira Oishi & Seisuke Miyazaki
Yachiyo Engineering Co., Ltd., Nishiochiai, Shinjyuku-ku, Tokyo, Japan

Naoki Kohara
Nippon Koei Co., Ltd, Kojimachi, Chiyoda-ku, Tokyo, Japan

ABSTRACT: It is important to understand river and groundwater interactions. The authors attempted to clarify these interactions by a survey of the hydrogeological structure of the study area. The alluvial fan sediments mainly consist of debris flow deposits. The gravel composition of surface outcrops includes granite on both banks of the R. Koishiwara-gawa, but these were not present on the left bank of R. Sata-gawa. The epiclastic sediments of the Aso-4 is divided into Aquifer I and II. Within the present annual water balance from 2003 to 2007, the recharge to groundwater was larger than the discharge from groundwater. It was estimated that the annual total surface flow was 400,000,000 m^3/day, and that the ratio of abstraction to the whole discharge was quite small. It was observed that groundwater closely interacts with river water, and that the recharge from the irrigation water amounts to about 40% of that from rainfall.

Keywords: Chikugo-gawa alluvial fan, Ryochiku Basin, hydrogeological structure

1 INTRODUCTION

The study area is located on the Ryochiku Basin in the northern Kyusyu district under non-snow coverage region where snowfall melts immediately, and the annual precipitation is about 1,500–1,900 mm. The Chikugo-gawa alluvial fan was formed by the rivers Koishiwara-gawa and Sata-gawa, which are flowing from the Kosyo-Umami Mountains of the north side of Ryochiku Basin (Fig. 1).

It seems that the groundwater environment of this area has been largely unaffected by human activities up to recent times, although reductions in spring discharges and decreases of the flowing wells have been caused by lowered groundwater level due to pumping from deep wells to serve the industrial and large-sized commercial buildings and by progressive urbanisation.

The Chikugo-gawa alluvial fan has been inhabited since the Yayoi Period, and was used for paddy fields which were irrigated from the spring waters and irrigation ponds on the alluvial fan until the 1950's as recorded in the Old Map of the Akizuki Clan created in 1842, shown in Fig. 2.

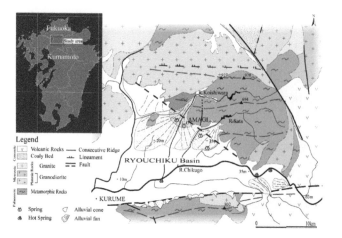

Figure 1.   Study area.

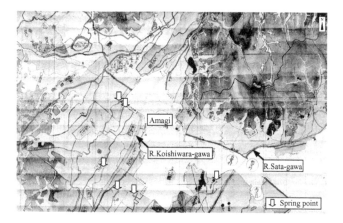

Figure 2.   "Akizuki-Funai-zu" published in 1842.

Cropland abandonment and urbanization increased recently, although the arrangement of the irrigation networks and the arable land holdings were upgraded by the construction of Egawa Dam and Terauchi Dam and were made into rice fields in the 1960's.

In the R. Sata-gawa, "Segire" (stream loss) occurs during non-irrigation periods. Therefore the people in the villages are calling R. Sata-gawa the Japanese radish stream "Daikon-gawa". Because they cannot wash Japanese radish "Daikon" which fruited during late autumn to early spring, it is handed down as a "Kobo Daishi" tradition.

On the other hand, R. Kogane-gawa from where the spring water "Suizenji-nori" (a species of a blue algae) is cultivated. At present, spring cultivated. At present, spring water is not drawn except for the wet season, and the river is enriched by pumping from the wells.

Therefore, it is important to understand that the river and groundwater interact. The authors attempted to clarify these interactions by a survey of the hydrogeological structure of the area.

Geological techniques, measurement of the water balance and groundwater table, and water quality will be used for clarifying the hydrogeological structure.

The alluvial fan sediments were subdivided by Facies analysis of gravel beds. And the pyroclastic flow deposits were identified in the alluvial fan sediments. Basement rocks are Granitic rocks and Sangun Metamorphic rocks. There are some concealed faults near the pediment.

Based on these facts, the aquifer was classified. This hydrogeological model would account for water balance and water quality without inconsistency.

On the other hand, as mentioned in the IPCC Fourth Assessment Report, global warming has been steadily progressing. It is considered that global warming will causes an increase in evapotranspiration and changes in the hourly to daily rainfall patterns.

The question is how the groundwater environment will change in the study area as affected by similar phenomenon in the future. The past and future changes of the groundwater environment will be estimated and forecasted respectively using a numerical simulation model with simple hydrogeological models. The simulation result based on the hydrogeological model are described later.

2 STUDY AREA

2.1 *General*

The R. Chikugo-gawa is the largest river of Kyusyu Island. It originates in the Kuju volcano (1,787 m) and has a length of 143 km, reaching the Chikugo Plain through some intermountain basins. The catchments area is 2,860 km^2.

The Chikugo Plain, a productive grain producing region, covers an area of 620 km^2 and is divided into two areas near Kurume by relative relief. The upstream side is called the Ryochiku Basin. The Ryochiku Basin is surrounded by horst-mountain (700–900 m) which were formed by uplifting in the Quaternary period. R. Chikugo-gawa flows through the central part of the plain towards the west.

The Chikugo-gawa alluvial fan is formed by the R. Koishiwara-gawa and R. Sata-gawa, which are flowing from the Kosyo-Umami Mountains of the north side of Ryochiku Basin. The alluvial fan covers about 63 km^2. R. Chikugo-gawa is eroding the distal fan area.

2.2 *Rainfall*

Although the Saga meteorological gauging station is slightly beyond the study area, it has acquired relevant long-term annual rainfall data for more than a century as shown in Fig. 3.

Figure 3.   Long-term trend of rainfall.

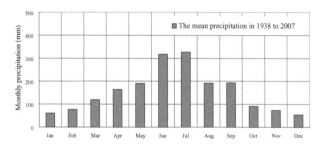

Figure 4.   The mean monthly precipitation for 1938 to 2007 (Saga meteorological station).

Over this period, monthly peak precipitation is in the rainy season (June to July); winter is the dry season (Fig. 4).

However, depending on the year, a peak can also occur during the typhoon season in autumn.

According to the Fig. 3, annual rainfall does not clearly show upward and/or downward trends, however, it reveals years of remarkable drought and high rainfall that frequently occurred especially during the last 30 years. This tendency, as indicated in Fig. 3, is also similar to the rainfall data from Asakura meteorological gauging station.

2.3   *Temperature*

As indicated in Fig. 5, temperature in Saga and Asakura clearly shows an upward trend especially after the 1950s, and the variation is estimated to be about 1.5°C for the last 20 years. However, it is known that land use change (especially due to urbanization) appears to influence the variability in daily maximum/minimum temperature. From this point of view, it can be considered that the slightly increasing daily minimum temperature in Saga is due to the fact that the surrounding area was fairly urbanized, especially after the 1980s.

On the contrary, the daily minimum temperature in Asakura does not show such a tendency. Hence, it is concluded that the recent warming trend in the study area is not caused by urbanization, but probably due to global warming.

2.4   *Stream flow of the main channel*

In the R. Chikugo-gawa catchment area, stream flow reflects seasonal changes in precipitation. It increases during the rainy season (June to July), then declines until winter (December to January).

Stream flow may increase in the short term during the typhoon season in autumn. Because winter snowfall rarely forms a continuous snow cover, the spring snowmelt does not increase stream flow.

Discharge of the R. Chikugo-gawa main stream reaches its peak in the rainy season at 0.13 m^3/s/km^2, then decreases to 0.02 m^3/s/km^2 in the dry season. Discharges of the R. Sata-gawa and R. Koishiwara-gawa are 30–80% and 20–70%, respectively, of that of the R. Chikugo-gawa (Fig. 6).

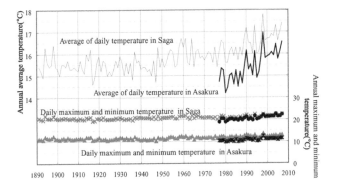

Figure 5.   Long-term trend of temperature.

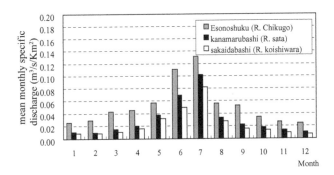

Figure 6.   Main river discharge.

3   GEOLOGICAL SETTING

3.1   *Geology*

The Cretaceous granites and the Triassic Sangun Metamorphic Rocks are distributed widely in the surrounding area of the Ryochiku Basin, and the volcanic rocks of Plio-Pleistocene age are distributed in the upstream area (Fig. 1). In the Ryochiku Basin, these rocks constitute an impermeable basement.

In the alluvial fan, two geomorphic surfaces (I: old, II: new) are identified, and springs are located on the Yorii-Amagi line at the distal fan area of alluvial-fan surface II. The lower terrace was 2–5 m below surface I along the downstream of the R. Koishiwara-gawa and R. Sata-gawa. A difference in level is clear at the distal fan area. On the left bank of the R. Sata-gawa, the former river-course surface I′ is lower than I (Fig. 7).

The Ryochiku Basin is a half graben rift basin formed by faulting in E-W and NW-SE directions. This basin is filled with debris flow deposits and flood plain to marsh sediments. These sediments have intercalated regional tephras, such as the Yufu-gawa pyroclastic flow deposit (hereafter Yfg) and Aso-4 pyroclastic flow deposit (Aso-4). The stratigraphy of the Chikugo-gawa alluvial fan is shown in Table 1.

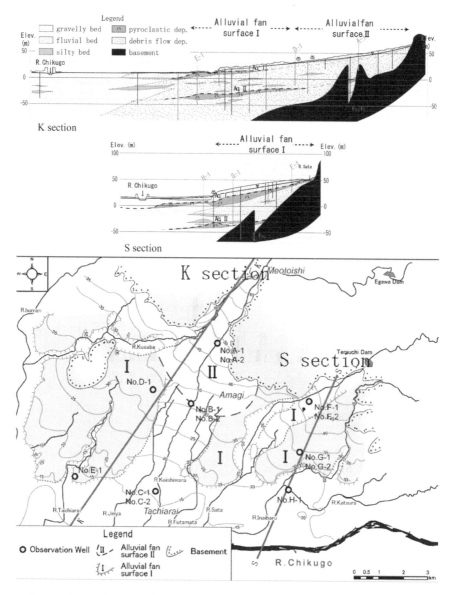

Figure 7.   Geological Profile of the Chikugo-gawa Alluvial Fan.

3.2   *Alluvial fan sediments*

The alluvial fan sediments mainly consist of debris flow deposits. The consolidated debris flow deposits are matrix-supported with thickness of 20–250 cm. Some sand beds with thickness of several centimetres occur in between along the boundary (Table 2, 1a). Grains are mostly sub-angular to angular gravel, and there is also a horizon in which rounded gravels are mixed. However, in alluvial-fan surface II, the sediments are loose compared to the lower debris flow deposits.

Table 1.   Stratigraphy of the Chikugo-gawa alluvial fan.

Cenozoic	Quaternary	Holocene	Present River Deposit
			Debris Flow Deposit
		Pleistocene	Aso-4 Pyroclastic Deposit (Aso-4, 90 ka)
			Debris Flow Deposit
			Flood Plane - Debris Flow Deposit
			Yufu-gawa Pyroclastic Deposit (Yfg, 600 ka)
			Flood Plane - Debris Flow Deposit
	Neogene	Pliocene	
Mesozoic	Cretaceous		Granitic rocks
	Triassic		Metamorphic rocks

Table 2.   Characteristics of typical facies.

Facies		Characteristics
Alluvial fan deposits		
1a	Debris flow deposit	Matrix support, Composed of sub-angular to angular gravel and matrix well compacted.
1b	Silt, Sand thin layer	With a 15 cm or less-thick granule mixture silt. Fine sand is at 3 cm or less in thickness. It is frequently inserted into a debris flow deposit.
1c	River-bed-gravel	It is distributed along the present riverbed. A high porosity clast-supported gravel bed
Flood plane deposits		
2a	Organic silt	Silt containing plant remains. Grey - dark grey. Contains sand particles.
2b	Medium-grained sand	Well sorted. It contains feldspar and colored minerals of volcanic-ash origin.
2c	Coarse sand Contained a rounded gravel	The coarse sand contains volcanic gravel. Schist gravel are not present.
Pyroclastic flow deposit		
3a	Yfg	Pumice and biotite are characteristic components. Well compacted. Grey color.
3b	Aso-4	Pumice and hornblende are characteristic components. It is a re-worked sediment Generally it is grey, with lamina of silt showing russet.

The gravel composition of the surface outcrops includes granite on both banks of the R. Koishiwara-gawa, that is hardly present on the left bank of the R. Sata-gawa (Fig. 8). In the fan surrounded by the R. Koishiwara-gawa and R. Sata-gawa, this figure shows that the sediment originated from the R. Koishiwara-gawa.

As it goes to the downstream lowland, flood plain silt and sand rate increase (1b). In the main channel, it is rich in pebbles and has high permeability (1c).

3.3   Sediments of the R. Chikugo-gawa lowland

In R. Chikugo-gawa lowlands, floodplain deposits consist of organic silt and well-sorted medium grained sand (2a, b), intercalated with coarse sand layers containing rounded volcanic rock fragments (2c) and schist. The former sand layer originated from the R. Chikugo-gawa up-stream area on the east, while the latter gravels of schist and volcanic rocks are originated from the upstream area of the alluvial fan.

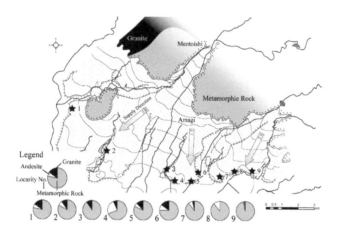

Figure 8.   Gravel composition and supply direction of alluvial fan deposits.

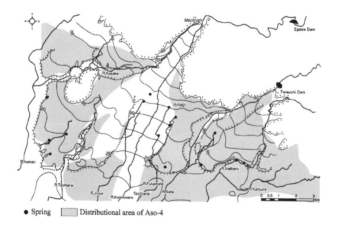

Figure 9.   Spatial distribution of Aso-4.

3.4 *Pyroclastic flow deposits*

Yfg is a grey pyroclastic flow deposit which erupted 80 km east of the R. Chikugo-gawa around 500,000 to 600,000 years ago, and has characteristic biotite crystals (3a). This pyroclastic flow deposit is distributed about 50–70 m below the surface of the Ryochiku Basin, although it is missing due to erosion beneath former and current river channels.

Aso-4 erupted 90,000 years ago, and it contains characteristic hornblende crystals (3b). The primary Aso-4 is dark grey with volcanic glass and pumice, and has a limited distribution. The epiclastic sediments from Aso-4 are rich in fine grained fragments with some grey-russet clay. Although Aso-4 is found 10–20 m below the surface on the right side of R. Koishiwara-gawa, it is lacking in some places (Fig. 9).

4 HYDROGEOLOGICAL FEATURES

The fan deposit is regarded as a cut and fill structure. According to the geological profile across the R. Koishiwara-gawa, the channel is filled up with the resedimental deposits. Moreover, according to the profile along the direction of dip, it is considered that the sediment of the fan surface II has covered the sediment of the fan surface I, at the proximal fan. The distribution of the slope of geomorphologic features and spring-water is a possible reason, which is also evident from Fig. 6. The erosion scarp of the fan surface I from which it is distributed over both banks of the R.Koishiwara-gawa, can not be observed at the upstream of Amagi.

Aso-4 was deposited before the last glacial stage and is not preserved in its primary form in the alluvial fan, although traces of it can be seen clearly on the fan surface. These traces are secondary deposits of the tephra. The surface weathers by a fine-grained clay-forming process, and a difference in hydraulic head is observed in the upper and lower aquifers. In this way, the upper aquifer has become Aquifer I, and the lower, bordering on Aso-4, is classified as Aquifer II. However, along the R. Koishiwara-gawa and R. Tachiarai-gawa, Aso-4 has often been reduced by erosion.

The basement rock, which consists of schist, appears in the upstream side of the alluvial fan, and is considered to be impermeable. On the downstream side and along the R.Chikugo-gawa, the basement rock cannot be confirmed even at a depth of 70 m. Although Yfg lacks continuity, it has low permeability compared with a debris flow deposit. Yfg is considered as the impermeable basement because it frequently contains silt particles.

5 PERMEABILITY

The permeability of each layer is summarized in Table 3, based on the permeability tests at the time of drilling the groundwater monitoring well by the MLIT (Ministry of Land, Infrastructure, Transport and Tourism).

The permeability test value of the Aquifer I is $2.41 \times 10^{-5} - 7.17 \times 10^{-4}$ m/s with an average of 3.62×10^{-4} m/s. The test value of the Aquifer II is $1.97 \times 10^{-6} - 1.00 \times 10^{-5}$ m/s with an average of 2.04×10^{-5} m/s, and is taken as $1.0 - 5.0 \times 10^{-5}$ m/s. The two aquifers are similar in their sedimentary facies, but differ in their permeability by one order of magnitude. Aquifer II may be in a compacted weathered debris flow deposit.

Table 3. Permeability and characteristics of aquifer.

| Aquifer division | Hydraulic conductivity (m/s) | | Facies | Water quality | | Ion concentration |
	Adoption value	Test value		EC (mS/m)	Type	
Present river deposit	5×10^{-3}	–	Grabel deficient in a matrix	8.5~20.0	Ca-HCO$_3$	
First aquifer	$3{\sim}9 \times 10^{-4}$	$2.41 \times 10^{-3} \sim 7.17 \times 10^{-4}$ (Average: 3.62×10^{-4})	Debris flow dep.	12.5~14.5	Ca-HCO$_3$	Low
Aso-4 pyroclastic flow dep.	$1 \times 10^{-6{\sim}7}$	–	Fine-grained secondary sediment, surface are clayization by a weathering	–	–	
Second aquifer	$1{\sim}5 \times 10^{-5}$	$1.97 \times 10^{-6} \sim 1.00 \times 10^{-4}$ (Average: 2.04×10^{-5})	Debris flow dep., well compacted	12.1~32.6	Ca, Mg-HCO$_3$ Ca-HCO$_3$	
Flood-plain dep. of R.Chikugo-gava	$1{\sim}5 \times 10^{-4}$	$3.80 \times 10^{-5} \sim 7.45 \times 10^{-5}$ (Average: 5.63×10^{-5})	Organic silt, River bed gravel	18.0	Na-HCO$_3$	
Yufugawa pyroclastic flow dep.	5×10^{-6}	$1.40 \times 10^{-6} \sim 7.65 \times 10^{-6}$ (Average: 4.81×10^{-6})	Fine-grained secondary sediment, Organic silt	25.7~30.0	Ca-HCO$_3$	
Metamorphic rocks	5×10^{-6} under	–	Pelitic Schist	–	–	High

The flood plain deposit of the R. Chikugo-gawa lowland has tested value of $3.80 \times 10^{-5} - 7.45 \times 10^{-5}$ m/s and an average of 5.63×10^{-5} m/s.

Although the test value has not been determined in Aso-4, about $1 \times 10^{-6 \sim -7}$ m/s is presumed from the characteristics. Because the test value in Yfg is $1.40 \times 10^{-6} - 7.65 \times 10^{-6}$ m/s with an average of 4.81×10^{-6} m/s, it is considered to be 5×10^{-6} m/s. The metamorphic rock has few cracks, and has a lower permeability than Yfg.

Although the test value has not been recorded from the present river deposit materials, it is presumed to be about 5×10^{-3} m/s from the characteristics of the material.

6   GROUNDWATER FLOW SYSTEM

6.1   *Groundwater table*

There are many shallow wells in the fan. We investigated wells where water sampling was possible. Measurements were carried out in April and November 2007 during the periods without irrigation, and in August during the irrigation period. These measurements showed that the profile of the groundwater table of the Aquifer I matches the geomorphic surface (Fig. 10). On the fan's surface, the geomorphic surface has a ridge on the downstream side and forms a valley along the river channel.

The R. Koishiwara-gawa is regarded as a typical alluvial fan river with characteristic continuously decreasing stream flows. On the fan's surface II which extends from Meotoishi to Amagi town, the profile of the groundwater table assumes a ridge downstream. A group of springs is distributed over Amagi town. At present, artesian flow from these springs is rare, even during the wet season. Even in August, groundwater level is tens of centimetres below the surface, and in April it falls to 3–7 m below the surface. Groundwater level in 1961 was higher than in August 2006, and it seems that artesian flow occurred in 1961 (Fig. 11).

The profile of the groundwater table of an alluvial fan extends to the downstream side as a ridge, and along an active channel it becomes valley. R. Tachiarai-gawa emerges from near the centre of the alluvial fan, dissecting the alluvial fan, and it is a river in which stream flow increases.

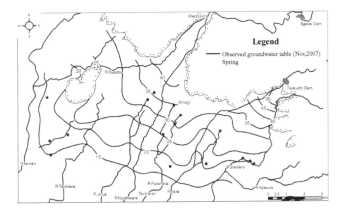

Figure 10.   Groundwater table on 25 November 2007.

On the left bank of the R. Sata-gawa, the groundwater table ridge runs through the central part, on the east side, and groundwater flows to the south and gushes in the erosion scarp of the fan toe. The former riverbed and the valley of the groundwater table are in agreement, and it has become a cause of stream lost "Segire". That is, stream flow decreases on the upstream side rather than in the mid-fan part.

In these sections, groundwater recharged the flows along the former channel and returns to the R. Sata-gawa after a while.

R. Kogane-gawa from where the spring water originates, "Suizenjinori" (a species of blue algae) is cultivated. At present, spring water is not drawn except for the wet season, and the river is augmented by pumping from the well.

A significant difference in the flow direction of groundwater was not observed in April and August. The flow in 1961 was also the same.

6.2 *Fluctuation in groundwater level*

In order to enumerate the seasonal fluctuation of groundwater levels, observation wells were newly installed at eight places (Fig. 7). Five of these were twin observation wells so that the groundwater level of Aquifer I and Aquifer II could be measured separately. In an observation well, groundwater level of the Aquifer I is always higher than that of the Aquifer II (Fig. 12). Groundwater is shown to have permeated below.

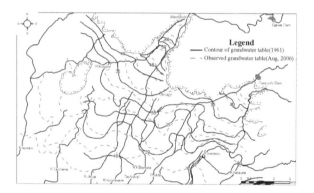

Figure 11.   Comparison of groundwater table on 1961 & August, 2006.

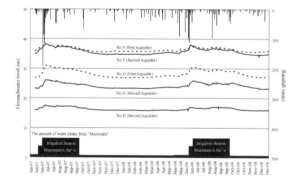

Figure 12.   Fluctuation in groundwater level (Left bank of R. Sata-gawa).

7 GROUNDWATER QUALITY

The groundwater, sampled from the observation wells, was analyzed in July and December to take account of seasonal changes. The analysis results are shown in Fig. 13 by hexagonal diagram of the main elements. The water type of Aquifer I is the Ca-HCO$_3$-type and it is similar to river water. Total ion concentration of groundwater from Aquifer II increases with varied Mg and Na ion concentrations.

On the vertical profile shown in Fig. 14, concentrations of Ca and NO$_3$ ions in Aquifer I increase from upstream to downstream. NO$_3$-enrichment may come from cultivated soils. As in Aquifer II, Ca and HCO$_3$ ions increase.

These tendencies will be stationary in July and December. Therefore, that is the phenomenon throughout the year without irrigation effects.

According to these observations, we assume that river water recharges Aquifer I each year.

8 SIMULATION

8.1 *Simulation of the water table*

The authors discussed the three-dimensional groundwater flow model based on the hydrogeological model. The initial hydraulic permeability was taken from the in-situ data, and after its validation, is shown in Table 4. In order to simulate the in-situ hydrogeological condition, the hydraulic permeability in the vertical direction of the pyroclastic flow deposit (Aso-4) was set as 1/10 of the hydraulic permeability in the horizontal direction.

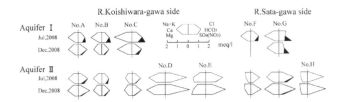

Figure 13.   The water quality of a groundwater.

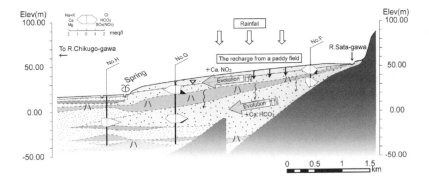

Figure 14.   Vertical profile on the R. Sata-gawa side.

Table 4.   Hydraulic conductivity after validation.

Layer	Hydrogeological name	Initial hydraulic permeability coefficient (m/sec)	Validated hydraulic permeability coefficient (m/sec)
1	River bed deposit	$5.0E^{-03}$	$5.0E^{-03}$
2	Aquifer I	$3.6E^{-04}$	$7.2E^{-04}$
3	Pyroclastic flow deposit (Aso-4)	$1.0E^{-06}$	Horizontal: $1.0E^{-6}$ Vertical: $1.0E^{-7}$
4	Aquifer II	$2.0E^{-05}$	$1.0E^{-04}$
5	Flood deposit	$5.6E^{-05}$	$2.8E^{-04}$
6	Debris flow deposit	$5.0E^{-06}$	$1.0E^{-05}$
7	Basement	$4.8E^{-08}$	$4.8E^{-08}$

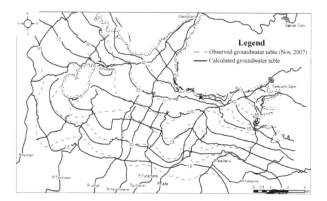

Figure 15.   Comparison of groundwater table on November, 2007.

The simulated water table represents the observed well water table, excluding those in the western area where the topography varies widely as shown in Fig. 15.

In general, the simulated groundwater level fluctuation also represents the observed groundwater level fluctuation in each observation station as illustrated in Fig. 16.

As for observation station G, which is located in the vicinity of the R. Sata-gawa course, there exists an approximate 5 m difference in potential head between the shallow and deep groundwater across an aquiclude (Aso-4). The potential head difference during the irrigation period is more than during the non-irrigation period. This difference in potential head can be approximately determined from the simulation, with a 2 m difference between the model and observations.

8.2   Present water balance

The present annual water balance from 2003 to 2007, simulated using the three-dimensional groundwater flow model, is shown in Fig. 17. During this period, the recharge to groundwater was larger than the discharge from groundwater.

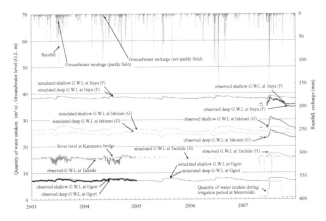

Figure 16.   Comparison between the simulations and observations along the R. Sata-gawa.

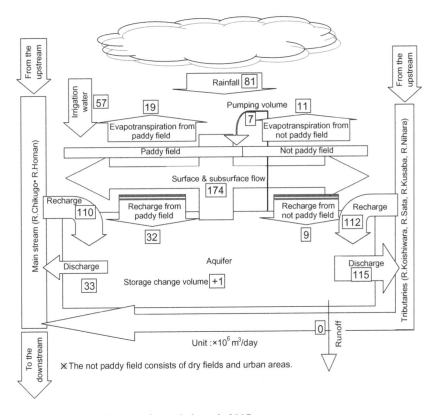

Figure 17.   Conceptual diagram of water balance in 2007.

It was estimated that the annual total surface flow was 400,000,000 m³/day, and the ratio of abstraction to the whole discharge was quite small. Meanwhile, it can be concluded that the water balance is adequately stable since the error (Σinflow–Σoutflow) is less than 0.1% for each year.

It was observed that groundwater closely interacts with river water, and that the recharge from the irrigation water amounts to about 40% of the recharge from rainfall as shown in Fig. 17.

9   SUMMARY

The summary of the findings of our survey on the groundwater flow system in the Chikugo-gawa alluvial fan is as follows.

1. The Chikugo-gawa alluvial fan is a compound alluvial fan formed by the R. Koishiwara-gawa and the R. Sata-gawa. Two old and new alluvial fans exist and spring water and ill-drained paddy fields are scattered at the edge of the fan on alluvial fan I of the old stage. There are many lotus paddy fields at the lowlands under the cliff in the distal fan part as well. Moreover, the Amagi city area is located in the vicinity of the spring water belt at the edge of the fan on alluvial fan II of the new stage.

2. The alluvial fan sediments mainly consist of debris flow deposits. The gravel composition of surface outcrops includes granite on both banks of the R. Koishiwara-gawa, but not on the left bank of R. Sata-gawa. In the fan surrounded by the R. Koishiwara-gawa and R. Sata-gawa, this shows that the sediment originated from the R. Koishiwara-gawa. On the downstream side, thin silt and sand that often form flood plain deposits are distributed at the mid-fan area.

3. The alluvial fan deposit is regarded as a cut and fill structure. Aso-4 is not preserved in its primary form in the alluvial fan, although traces of it can be seen clearly on the fan surface. The surface weathers by a fine-grained clay-forming process, and a difference in hydraulic head is observed in the upper and lower aquifers. In this way, the upper aquifer has become Aquifer I, and the lower, bordering on Aso-4, is classified as Aquifer II. However, along the R. Koishiwara-gawa and R. Tachiarai-gawa, Aso-4 has often been reduced by erosion.

4. The basement rock, which consists of schist, appears in the upstream side of the alluvial fan, and it is considered to be impermeable. Although Yfg lacks continuity, it has a low permeability compared with a debris flow deposit. Yfg is considered as impermeable basement because it frequently contains silt particles.

5. Groundwater table measurements showed that the profile of the groundwater table of the Aquifer I matches the geomorphic surface. The R. Koishiwara-gawa is regarded as a typical alluvial fan river with characteristic continuously decreasing streams flow. On the fan's surface II, which extends from Meotoishi to Amagi town, the profile of the groundwater table assumes a ridge downstream. A group of springs is distributed over Amagi town. At present, artesian flow from these springs is rare, even during the wet season. Even in August, groundwater level is tens of centimetres below the surface, and in April it falls to 3–7 m below the surface.

6. On left bank of R. Sata-gawa, the groundwater table ridge runs through the central part, on the east side, and groundwater flows to the south and discharges from the erosion scarp of the fan toe. The former riverbed and the valley of the groundwater table are in agreement, and it has become a cause of stream lost "Segire". That is, stream flow decreases on the upstream side rather than in the mid-fan part. Groundwater recharged in these sections flows along the former channel and returns to the R. Sata-gawa after a while.

7. In an observation well, the groundwater level in the Aquifer I is always higher than that of the Aquifer II. Groundwater is shown to have permeated below. The results of groundwater analysis suggest that river water recharges Aquifer I each year.

8. As compared to the groundwater contour map in 1961, the present groundwater table is lowered by about 2–3 m near the mid-fan. This explains the recent drying-up of the groundwater, and is also evident from the decrease of spring water and flowing wells. A decrease in the amount of groundwater recharge by the advancement of urbanization and an increase in the amount of groundwater abstraction are thought to be the reason.

9. The authors discussed the three-dimensional groundwater flow model based on the hydrogeological model. The simulated water table represents the observed water table, excluding those in the western area where the topography varies widely. In general, the simulated groundwater level fluctuations also represents the observed groundwater level fluctuations in each observation station.

10. In the present annual water balance from 2003 to 2007, the recharge to groundwater was larger than the discharge from groundwater. It was estimated that the annual total surface flow was 400,000,000 m^3/day, and the ratio of abstraction to the whole discharge was quite small. It was observed that groundwater closely interacts with river water, and that the recharge from the irrigation water amounts to about 40% of the recharge from rainfall.

Generally, due to the rise in temperature in the future, the amount of evapotranspiration and river outflow will increase due to the change in the rainfall patterns and the amount of recharge to groundwater is expected to decrease. Additionally, further demand for 'fresh water' is expected due to improvements in human quality of life and the resultant changes in the industrial structure.

Therefore, it is important to understand that the river and the groundwater have interacted with one another. And, it is advisable to manage the river water and the groundwater in a unified concerted manner in order to utilize effectively the groundwater, which is a stable freshwater resource, after taking into consideration the hydrogeological structure of the Chikugo-gawa alluvial fan and the groundwater flow characteristics in this region.

ACKNOWLEDGEMENT

The authors wish to thank all the parties concerned with the Chikugo-gawa River Office, Kyusyu Regional Development Bureau, MLIT, the Ryochiku Basin Water-for synthesis place of business, JWA and Ryochiku Land use office, Fukuoka Prefecture. They helped with observation of core samples, analysis of the results of the permeability tests and a survey of the inspection wells, for the study of the water environment of the Chikugo-gawa alluvial fan.

REFERENCES

Amagi City (1982) Regional Geography of Amagi, First Volume, the Compilation commission of the history of Amagi.
Amakata, M. et al. (2006) Investigation and Research for Qualitative Understanding of Ground Water Trend in the Ryochiku Basin. Journal of Japan Society of Hydrology and Water Resources, 19(1), 61–66.

Ariake Bay Research Group (1969) Quaternary of Kyusyu area. Association for the Geological Collaboration in Japan new report, Quaternary of Japan, Vol. 15, 411–427.

Fukuoka Prefectural Sabo Association (2005) 1:150,000 Geological map of Fukuoka Prefecture.

Hasegawa, S., Takada, K., Shimada, J., Shimoosako, H., Research Group on Hydro-environment Around Alluvial Fans (2006) Study of Alluvial Fan (Part 8) The Groundwater Flow System in Alluvial Fan, Chikugo Area (Preliminary Report). Proceedings of Meeting, Japan Society of Engineering Geology, 157–160.

IPCC WG II 4th Assessment Report (2007) Climate Change Impacts, Adaptation and Vulnerability. 987p.

Kido, M. (1997) Fault Development of Minoh Range and the Kitano Plain, Northern Kyusyu, Southwest Japan. Journal of the Geological Society of Japan, 103, No. 5, 447–462.

Kuroda, K., Kuroki, T. (2004) Landform Development at the Northern Part of Kitano Plain after Aso 4 Pyroclastic Flow Deposition. Proceedings of the General Meeting of the Association of Japanese Geographers, 65, 81–81.

Kuroda, K., Kuroki, T., Kagashima, S. (2004) Development of Terrace at the Northern Part of Kitano Plain after Aso 4 Pyroclastic Flow Deposition. Program and Abstracts, Japan Association for Quaternary Research, 34, 111–112.

Kuroki, T., Kuroda, K., Nakamura, Y. (2003) Relationships between Characteristics of the 1953 Flood Disaster and Microtopography in the Kitano Plain. Proceedings of Meeting, Japan Society of Engineering Geology, 267–270.

Machida, H., Arai, F. (2003) Atlas of Tephra in around Japan. University of Tokyo Press.

Matsumoto, T., Miyazaki, S., Oishi, A., Research Group on Hydro-environment Around Alluvial Fans (2006) Study of Alluvial Fan (part 7) Geomorphology and Geology of Alluvial Fans in the Chikugo Area. Proceedings of Meeting, Japan Society of Engineering Geology, 153–156.

Author index

Subject index

SERIES IAH-Selected Papers